Media and Global Climate Knowledge

Risto Kunelius • Elisabeth Eide • Matthew Tegelberg • Dmitry Yagodin
Editors

Media and Global Climate Knowledge

Journalism and the IPCC

Editors
Risto Kunelius
Department of Social Sciences
University of Tampere
Tampere, Finland

Elisabeth Eide
Journalism and Media Studies
Olso & Åkerhus University College
Oslo, Norway

Matthew Tegelberg
York University
Toronto, Ontario, Canada

Dmitry Yagodin
School of Communication,
Media and Theatre
University of Tampere
Tampere, Finland

ISBN 978-1-137-52320-4   ISBN 978-1-137-52321-1   (eBook)
DOI 10.1057/978-1-137-52321-1

Library of Congress Control Number: 2016957872

Cover illustration © Elisabeth Eide, from Cox's Bazar, Bangladesh, January 2013

Printed on acid-free paper

This Palgrave Macmillan imprint is published by Springer Nature
The registered company is Nature America Inc. New York
The registered company address is: 1 New York Plaza, New York, NY 10004, U.S.A.

FOREWORD

This book would not have been possible without an enduring network collaboration between researchers in more than 20 countries. The *MediaClimate* network started its work after the UNFCCC Bali summit in 2007 (COP 13), with the firm conviction that climate change is one of the major global challenges of our time. We believe that journalism, in all its variety of forms and outlets, plays an important role in informing and engaging citizens faced with this challenge. Journalism is also crucial in building the public pressure to help decision-makers shape urgently needed policies. This is much more than a normal challenge of spreading and popularizing complex scientific knowledge. As we argue in this book, climate change digs deep into many, if not all, areas of social and political life. Thus, journalism must not only follow the science but also demonstrate the interrelatedness between climate change and other major challenges, such as migration, financial crisis and the unequal distribution of wealth.

This book takes the *Intergovernmental Panel on Climate Change's Fifth Assessment Report* (IPCC AR5) as its point of departure. The report was published in four parts between September 2013 and November 2014. This extended event offered a unique opportunity to take stock of the science–journalism relationship in all its complexity. To underline the importance of the results and the gravity of the situation, the IPCC organized a conference (Our Common Future) in Paris in July 2015. At this event, leading scientists highlighted their findings and—albeit sometimes between the lines—urged politicians and other stakeholders to take them seriously. Loaded with potential political impact, the AR5 report was released in due time for the world leaders who gathered in Paris for COP 21 in December 2015. These events were unique moments for our team to study media content and interview scientists and journalists concerning their views on the challenges of communicating climate science. We are grateful to the IPCC scientists and reporters who found time to share their valuable experiences and insights with us during busy days in Paris, and at other meetings.

MediaClimate has been monitoring coverage of the global climate summits every second year from COP 13 in Bali (2007) to COP 21 in Paris (2015). During this time, the network has grown to cover over 20 countries across all continents. Our deepest gratitude goes to all those who have helped the authors of this volume by contributing their work. We have relied upon a large, skilful group of coders and colleagues who have brought their local knowledge and commitments to the project. The study would have been difficult to pursue without Amin Alhassan, Armèle Cloteau, Caroline D'Essen, Katherine Duarte, Lew Friedland, Kemantha Govender, Heli Heino, Daniel Hermann, Amanda Hines, Nissim Katz, Ville Kumpu, Zbyszek W. Kundzewicz, Wytold Kundzewicz, Emma Larsson, Heba Metwally, Mulatu Moges, Anja Naper, Elisa Ricciardone, Arin Sen and Henning Tegelberg.

Many aspects of this work have been presented at international conferences and seminars where numerous colleagues have provided important feedback and criticism. Their names would make an even longer list. We offer our collective thanks.

Over the years, the network has been generously supported by *Helsingin Sanomat Foundation* and the GIMD facility provided to Oslo and Akershus University College by the Norwegian Ministry of Foreign Affairs. The IPCC AR5 project, in particular, relied heavily on *Helsingin Sanomat* support. It has been a pleasure to work with the Foundation leaders, Heleena Savela and Ulla Koski. We would also like to acknowledge support from the Academy of Finland (Mediatization of Governance, 2015–2019). Finally, we would like to thank all of the universities whose academics have contributed to this network for providing the opportunity for our colleagues to take part in this endeavour.

June 2016   Oslo, Tampere and Toronto,
Elisabeth Eide, Risto Kunelius,
Matthew Tegelberg and Dmitry Yagodin

CONTENTS

Notes on the Contributors

Shinichiro Asayama (asayama.shinichiro@nies.go.jp) is a postdoctoral research fellow at the Center for Social and Environmental Systems Research, National Institute for Environmental Studies, Japan. His research focuses on discourses and framings of media, politics, science and technology in climate change coverage.

Philip Chubb (philip.chubb@monash.edu) is Head of Journalism at Monash University, Australia. His research interests include political and environmental communication. Last year he published a book *(Power Failure*, Black Inc., 2014) analyzing the inability of successive governments to implement emissions trading schemes.

Armèle Cloteau (armele.cloteau@gmail.com) is a PhD Candidate in the Department of Sociology and Political Science at Paris-Saclay University, France. Her research focuses on public policy at the French and European level, including the roles and media representations of industrial interest groups.

Jean-Baptiste Comby (jbay20@gmail.com) is an assistant professor at the French Press Institute, University of Paris 2, France. His research in sociology focuses on ecology and social transformation, social classes and environmental issues (Recent book: *La question climatique: Genèse et dépolitisation d'un problème public*, 2015).

Elisabeth Eide (Elisabeth.Eide@hioa.no) is Professor of Journalism Studies at Oslo and Akershus University College of Applied Sciences, Norway. She has led several research projects, written, edited and co-edited a number of books, including three on climate change and journalism and five novels.

Oliver Hahn (oliver.hahn@uni-passau.de) is a Professor of Journalism at the Centre for Media and Communication at the University of Passau, Germany. His research and teaching interests include international comparative journalism studies, science communication (media and climate change), media in South East Asia and in the Arab world and development journalism.

Li Ji (jliwhu@whu.edu.cn) is a professor in the School of Journalism and Communication at Wuhan University, China. Her research and teaching interests include intercultural communication, environmental communication and media culture.

Risto Kunelius (Risto.Kunelius@staff.uta.fi) is dean of the School of Social Sciences and Humanities at the University of Tampere, Finland. His research interests include media and power, mediatization and social theory, the changing role of journalism and public sphere theories.

Johan Lidberg (johan.lidberg@monash.edu) is a senior lecturer in the School of Media, Film and Journalism at Monash University, Australia. His research interests include information access systems (freedom of information), climate change communication and media accountability.

Débora Medeiros (deb.medeiros@fu-berlin.de) is a PhD candidate in Communication and Media Studies at the Free University of Berlin. Her current research focuses on media practices in alternative coverage of protests. Previous research topics include citizen media and environmental reporting, as well as public service broadcasting.

Goretti Linda Nassanga (nassanga@chuss.mak.ac.ug) is an associate professor in the Department of Journalism and Communication at Makerere University Kampala, Uganda. Her areas of research expertise include development communication, gender and media, community media, environment/climate communication, international communication, population and health communication, new media and ICTs and peace journalism.

Hillel Nossek Ph.D. (hnossek@colman.ac.il) is a professor of Communication at Kinneret College on the Sea of Galilee, Israel. He is Chair of the Mediated Communication, Public Opinion and Society

Section of the International Association of Media and Communication Research (IAMCR). His main research interests and publications concentrate on foreign news and foreign news flow, media and political violence in general and media and terrorism in particular.

James Painter (james.painter@politics.ox.ac.uk) is Director of the Journalism Fellowship Programme at the Reuters Institute for the Study of Journalism, Oxford University, United Kingdom. His research interests include the portrayal of climate change in international media, the presence of skepticism, risk and uncertainty frames and the arrival of new players covering environmental issues.

Mofizur Rhaman (mofizur.rhaman@gmail.com) is a professor and chairperson of the Department of Mass Communication and Journalism at the University of Dhaka, Bangladesh. His research interests include environmental journalism, strategic communication, media representations of gender and intersectionality.

Anna Roosvall (anna.roosvall@ims.su.se) is an associate professor in the Department of Media Studies, Stockholm University, Sweden. Her research is centered on media and globalization in four related areas: media, climate justice and indigenous peoples; identity construction in world news images; reporting on mobile minorities and rights; and political-ethical dimensions in cultural journalism.

Adrienne Russell (adrienne.russell@du.edu) is an associate professor at the University of Denver with a joint appointment in the Emergent Digital Practices program and the Department of Media, Film & Journalism Studies. She is author of *Networked: A Contemporary History of News in Transition* (Polity, 2011) and *Journalism as Activism: Recoding Media Power* (Polity, 2016).

Ibrahim Saleh (ibrahim.saleh@uct.ac.za) is Convenor of Political Communication at Cairo University, Egypt. Saleh is Chair of the Journalism Research and Education Section at the IAMCR. His research interests include political communication, securitization of the environment, the impact of media violence on public opinion and digital inequality in cross-national contexts.

Billy K. Sarwono (billysarwono@gmail.com) is a professor in the Department of Communication Science in the Faculty of Social and Political Sciences, Universitas Indonesia. Her research and teaching interests include media representations of women and children and environmental communication.

Matthew Tegelberg (mtegel@yorku.ca) is an assistant professor in the Department of Social Science at York University, Canada. His research and teaching interests focus on global tourism, environmental communication and media representations of indigenous peoples.

Dmitry Yagodin (dmitry.yagodin@uta.fi) is a postdoctoral researcher in the School of Communication, Media and Theatre at University of Tampere, Finland. His research interest lies in the area of strategic communication, alternative media practices and digital media cultures.

LIST OF ABBREVIATIONS

ACC	Anthropogenic Climate Change
AR4	Fourth Assessment Report (2007)
AR5	Fifth Assessment Report (2013–14)
BRICS	Brazil, Russia, India, China, and South Africa (association of five emerging economies)
CO^2	Carbon Dioxide
COP	Conference of the Parties
COP 21	Paris Climate Summit
G77	Group of 77
GDP	Gross Domestic Product
GHGs	Greenhouse Gas Emissions
IPCC	Intergovernmental Panel on Climate Change
NGO	Non-Governmental Organization
SPM	Summary for Policy Makers
SYR	Synthesis Report (2014)
UN	United Nations
UNESCO	United Nations Educational, Scientific and Cultural Organization
UNFCCC	United Nations Framework Convention on Climate Change
WGI	Working Group I : The Physical Science Basis (2013)
WGII	Working Group II: Impacts, Adaptation, and Vulnerability (2014)
WGIII	Working Group III: Mitigation of Climate Change (2014)
WWF	World Wildlife Foundation

Note on IPCC AR5 abbreviation. Throughout this book IPCC AR5 is the abbreviation used to refer to the Intergovernmental Panel on Climate Change Fifth Assessment Report (2013/14). In the chapters, after initially referring to the IPCC AR5 the title is condensed further to AR5.

List of Figures

LIST OF TABLES

CHAPTER 1

The Problem: Climate Change, Politics and the Media

Risto Kunelius and Elisabeth Eide

The stakes are massive, the risks and uncertainties severe, the economics controversial, the science besieged, the politics bitter and complicated, the psychology puzzling, the impacts devastating, the interactions with other environmental and non-environmental issues running in many directions. The social problem-solving mechanisms we currently possess were not designed, and have not evolved to cope with anything like an interlinked set of severity, scale and complexity.

(Dryzek et al. 2011: p. 3). The Oxford Handbook of Climate Change and Society

[Climate change] can disperse power into the hands of the many rather than consolidating it in the hands of the few, and radically expand the commons, rather than auctioning it off in pieces. And where right-wing shock doctors exploit emergencies (both real and manufactured) in order

R. Kunelius (✉)
School of Social Sciences and Humanities, University of Tampere, Tampere, Finland

E. Eide
Department of Journalism and Media Studies, Oslo and Akershus University College of Applied Sciences, Oslo, Norway

© The Author(s) 2017
R. Kunelius et al. (eds.), *Media and Global Climate Knowledge*,
DOI 10.1057/978-1-137-52321-1_1

1

> *to push through policies that make us even more crisis prone, the kinds of*
> *transformations discussed in these pages would do the exact opposite: they*
> *would leave us with both a more habitable climate than the one we are*
> *headed for and a far more just economy than the one we have right now*
>
> (Klein 2014: p. 10). *This Changes Everything. Capitalism vs. The*
> *Climate*

The quotes above—one from a handbook by social scientists and one from an action manifesto by a radical global journalist and author—say the same thing in different ways. Climate change is a historically unique problem. Indeed, it is key proof that humans and their actions have become "central" on a planetary scale. This importance betrays a daunting sense of weakness. Perhaps we have become too central in our political and organizational capacities. Thus, paradoxically, from the overwhelming complexity of climate change as a challenge (in Dryzek), dawns a glimmer of radical system critique (in Klein): in order to adapt to what is coming, we must transform ourselves.

This book examines a specific moment in the ongoing global political process of trying to come to terms with what should be done: the publication of the Fifth Assessment Report of the Intergovernmental Panel on Climate Change (IPCC AR5). We focus on a particular aspect of this moment: the role of mainstream journalism in translating the "global knowledge" of the IPCC for local publics. By focusing on the launching moments of the four sub-reports of AR5 (during 2013 and 2014), we map how the media builds a discursive and social *space of interpretation* where the impacts of climate change are negotiated. By using the publication of the IPCC reports as an entry point to the flow of global, journalistic climate communication, we map some key aspects of the transnational communication infrastructure that is mobilized around global climate politics.

Three overlapping aspects of the climate change problem frame it as a communication challenge. First, there is the exceptional *scale*. Climate change saturates decision-making from everyday life and local politics to global governance; it penetrates deep into the structural conditions of modern societies and their social order and reaches from the distant past hundreds of years into the future. Hence, there is *no outside* position from which to formulate questions, construct knowledge or give advice. Every participant in the climate debate is part of the problem, an interested party. We take part in the discussion from our particular corners governed by the horizons and social rules of our own localities. The space of interpretation is, in this sense, fragmented and diverse. At the same time, we know there is another scale at work: the problem is a global one and thus demands a well-functioning dialogue that can transcend localities and help to negotiate and justify joint action.

Second, there is the *complexity*. It grows not only from the fact that the earth's climate system is itself a dizzyingly complex whole, but also from the fact that almost "everything" is or can be linked to climate politics (see Klein above). The debate about "climate" is never only about climate, but also about something else: energy, power, water, economy, land, employment, culture, tradition, identities, justice, survival, religion, food, mobility, lifestyle, democracy, freedom and so on. In this sense, our discussion about climate change resembles other "global discourses" such as debates about "human rights" or "free speech". Such universalities are always articulated through particularities. Imagining solutions calls for translation between different kinds of knowledge and vocabularies. This underlines the importance and difficulties of communication.

Third, the combination of scale and complexity becomes sharply articulated as a *political* and democratic challenge. Climate change extends the limits of the narrow jurisdiction of political institutions. It provokes anxiety about a fundamental temporal mismatch between climate policy and electoral democracy as a way of coordinating legitimate action. It raises fundamental questions about what democratic politics is and should be. Since the media is fundamentally paired up with our democratic imagination, these tensions also make themselves felt, particularly when we talk about journalism.

In order to understand media and journalism, it is necessary to look at the broad context in which it is situated. Hence, we anchor our analysis of *climate* journalism, by first briefly considering the three aspects introduced above: scale, complexity and (democratic) politics.

Scale: Width, Depth and Time

Climate change is an exceptionally *wide* challenge. The emission of greenhouse gases affects all humans and other living beings, the dynamics of the earth's climatic system and the entire planetary ecosystem. Greenhouse gases travel easily and have consequences far beyond their place and time of origin. This underlines the interdependency of peoples, localities and generations. Global warming effects transgress the borders of modern political communities (mostly nation-states) that make up the fragile system of international governance. Climate is not the first nor the only global problem that highlights the "Post-Westphalian" or "Post-National" conjuncture in which we find ourselves (Habermas 2001, 1992; Fraser 2007, 2014). However, historically, it is arguably the most difficult and complex problem that humanity as a whole has faced. As recent attempts to write "global history" show (Mann 2013; McNeil and McNeil 2003; Fernández-Armesto 2011), climate change appears as perhaps the test case for the emerging twenty-first-century globalized system of power. History teaches us that so far rescaling systems of governance to the level of inclusive global problem-solving has never been done effectively. Success on this scale entails formidable communication challenges, not only in terms of constructing broadly shared realities concerning what we know about climate change but also in terms of what would be a reasonable and justified path to react to it.

Climate change also poses an exceptionally *deep* problem regarding the base structure of our social system. Its roots are our roots, the "solid ground" on which the dominant economic, political, cultural and social order of our world is based. Modern societies are "carbon-thirsty" by definition (Urry 2010) with structures of power intimately linked to the ways they have organized themselves around energy solutions. As economic growth curves, measures of material well-being or human mobility (social and geographical) have pointed upward, the roots of the fossil fuel economy have dug deeper into the ground on which we stand. Whether a shift to alternative sources of energy can provide the grounds for further "growth" in the current system of global capitalism remains uncertain. Ignored in the past, the economic "externalities" of pumping carbon into the atmosphere are forcing themselves back into our calculations. Doing so has proved difficult and wrought with political disagreements. The dominant economic debates mostly limit themselves to arguments concerning what big investments would be "rational" in a climate change cost-benefit analysis. (Stern 2007; Nordhaus 2008; Piketty 2014; Dietz and Stern 2014). Such calculations naturally assume that the system must remain in place (i.e. the current, albeit messy, relationship between states, politics and the economy) and that we can move from carbon to other energy sources quickly and effectively (i.e. in a matter of decades). However, the emission curves still point alarmingly upward. Furthermore, we know that cost-benefit models of economics tend to factor into their calculations values that "...are rooted in the economy we are trying to leave", as Dryzek et al. (2011: p. 46) point out. Putting this more clearly, Malm and Hornborg (2014) argue that "...uneven distribution is a condition for *the very existence* of modern, fossil-fuel technology" (p. 3). Sagoff (2011) suggests that climate change is not a "market problem" at all, as it may be unsolvable within the language and conceptualizations of market economists. These critiques point to the depth of the climate challenge and to the narrow limits of mainstream economic calculus in addressing problems with system-level catastrophic risks.

One key instability in economic assessments reflects yet another scale dimension of the climate problem: the exceptional *time* horizons of climate politics. In economic disputes this circulates around estimations of the proper "discount rate" with which future hazards and benefits are brought forward in today's choices. But climate change also shakes our timescales in more fundamental ways. Dominant public identities in modern societies are characterized by a sense of being situated in a time of progress, in a continuum that stretches from pre-modern to late modern societies—and sometimes beyond. Strong narrative tropes in our cultural imagination present this continuum as a story of growth that continuously generates higher levels of development. At the same time, a variety of counter-narratives challenge this naturalized story of "development by capitalist growth". Climate change has activated both of these cultural plots, provoking apocalyptic predictions and fantasies about engineering the future of nature (Hulme 2010, 2014), demonstrating the pervasive sense of both *teleological* time and fatalism inside modernity. Yet debating climate change demands a multi-timescale view of politics. On one end there is a need to relate to extremely long developments in the climatic system. Can we actually "make sense" of what hundreds of thousands of years mean? Do we actually care about life (some) hundred years from now? An investigation of laypeople's views on future socio-economic and climatic change in the UK and Italy revealed that that participants "…felt 50 years to be a very long time-span for personal visioning. Imagining long-term futures also proved difficult as the quickening pace of (societal) change over the last 50 years meant the past could not be relied upon as a yardstick for future change" (Lorenzoni and Hulme 2009: p. 391). The knowledge produced by the IPCC is itself a good example of the complexity of simultaneous timescales, arguing at the same time on horizons such as "the last 800,000 years", "since the pre-industrial time" (about 250 years) and the "space for solution" closing in 2030 or 2050 (15–35 years).

Each of these dimensions—width, depth and time horizon—of the climate challenge, then, call for *recognition* of the all-embracing universal scale of the problem. A credible and effective solution to the fast mitigation challenges by means of a global agreement demands a binding inclusive target and a fair "deal" with no free riders. Thus, a plausible and sustainable solution to move beyond global, carbon-thirsty capitalist modernity would call for fundamental re-considerations of identities, economic and social orders, and political practices. At the same time, action toward these goals must be designed in a world where such boundary crossing arguments or agreements do not command universal obedience. This framework sets the stage for thinking about how to communicate climate science and climate politics. How will journalism be able to report the science and its predictions to differently situated publics around the world? How can it critically cover transnational negotiations and international governance efforts, while enhancing their necessary authority? How are the voices of people in radically different situations, locations and time horizons to be included in the debate that informs such governance?

COMPLEXITY: KNOWLEDGE, CIVIC EPISTEMOLOGY, INSTITUTIONS, INEQUALITY

The earth's climate system is itself a complicated, dynamically balanced entity. This complexity makes climate change an illustrative case of an object of modern science where the things *not* known and the institutionalized practice of *doubt* are constitutive characteristics. Considering this legacy of scientific doubt and the complexity of the object, the volume of multi-disciplinary knowledge collected and *synthesized* by the IPCC is a remarkable achievement. For instance, the Synthesis Report (December 2014) reaches an increasingly conclusive level of language:

> Anthropogenic greenhouse emissions have increased since the pre-industrial era, driven largely by economic and population growth, and are now higher than ever. This has led to atmospheric concentrations of carbon dioxide, methane and nitrous oxide that are unprecedented in at least the last 800,000 years. Their effects, together with those of other anthropogenic drivers, detected throughout the climate system and are extremely likely to have been the dominant cause of the observed warming since the mid-twentieth century. (IPCC 2014: p. 4)

"Extremely likely", in the technical language of the IPCC, translates to 95 to 100 percent certainty. Here, we see complexity turned into clarity. There *is* a clear enough causality, a practical certainty concerning the core mechanism and causes of warming. There is a correlation between emissions and the rise of global temperature. We know that we have already spent 65 percent of the "carbon budget" at our disposal if we are to have a reasonable chance of staying under two degrees of average global temperature rise. Considering the epistemological culture of doubt that drives modern science, there is little chance of that verdict getting any more conclusive.

However, the complexity of the problem quickly returns when we try moving from these synthesized findings to practical responses. First, complexity reappears in the diverse nature of knowledge itself. Hulme (2010) has argued, for instance, that the "global knowledge" produced by the IPCC is often too abstract (e.g. the *average* two degree temperature limit). It is also based on globalizing particular environmental values (such as the *economic* cost-benefit language). Thus, while recognizing the need to address climate change with such globalized models and measures (to at least raise attention), Hulme (2010) warns against knowledge that is "...insensitive to the peculiarities of place and context, and that "...opens the way for unitary globalized explanations and predictions of environmental change" (p. 559). Against such evidence "...masquerading as universal truths [that] assert themselves as the unassailable view from everywhere", he subscribes to a more "cosmopolitan view". From this perspective, developed in more detail by Beck (2010), Hulme (2010) argues:

> Rather than seeking a consensual global knowledge which erases difference and allows the most powerful to determine what is "known", we need to pay greater attention to the different ways knowledge comes to be made in different places and how different kinds of knowledge gain hold in people's minds, traction in different cultures and assent in global fora. This is *spectral knowledge* which emerges from a cosmopolitan perspective. (p. 563; emphasis added)

The call for spectral knowledge underlines the very locality of knowledge itself. It also begs the question to what extent (and in what sense) knowledge can be detached from its context of use and production. Thus, beyond the complexity of the physical phenomena (that the IPCC captures, synthesizes and simplifies) emerges another complexity of cultural, social and local conditions and experiences. For Hulme, this is not merely a matter of the complexity of "applying" the same global knowledge, but that the world consists of independent "ways of knowing". In other words, there is unresolved complexity at the heart of the very knowledge that the IPCC is to trying synthesize. For our study of journalism and climate communication, this presents a huge challenge. How can climate science be mediated in a way that takes heed of the dire, globally necessary abstract warnings but at the same time makes room for local knowledge and life experiences?

Second, diversity appears as a *plurality of civic epistemologies*, the varying political cultures, traditions and institutional arrangements that go into shaping local policy. Even if scientists could reach a unanimous, accurate, final global agreement on the facts—a unity is counter-intuitive to contemporary science—it would not translate into policy in a uniform manner since global climate facts are *domesticated* by local institutions and networks of power. The interpretation processes that are mobilized to craft policy vary according to political traditions, earlier institution-building and contemporary struggles. Jasanoff (2011), for instance, has pointed to such variations of governance by comparing the USA, the UK and Germany. She illustrates how the same message from science in different local "civic epistemologies" has resulted in different landscapes for climate policy. The common law culture of the USA—that sees the "truth" emerging in an adversarial courtroom style encounter—has produced polarized debates and has largely resulted in national inaction. The UK culture of empiricism has been more favorable to scientific authority, focusing on science and evidence but also remaining open to climate skepticism and denial. The consensus minded political culture of Germany, in turn, has supported more coherent political action, at least in the short run. One needs only to point out that these three countries are situated *within* a broad and largely shared Western legacy of democratic imaginaries to realize how much more local or national diversity the *global* context might mean. Whatever joint actions are called for in global climate politics, they must be crafted into national and local policies in political and administrative machineries that vary in structure and effectiveness as well as in the traditions of argumentation and the production of legitimacy and consent. Third, the complexity of the climate challenge is amplified by *the intensified and porous boundaries between* modern institutions of governance. If civic epistemology refers to the inherited local varieties of institutional configurations for crafting policy, the boundary issue highlights another example of how climate change forces us to rethink our modern legacy. Here at stake is the constitutive self-image of modern societies as social systems made of differentiated, autonomous sub-systems (science, politics, journalism, religion, etc.) with functionally distributed capacities and interdependencies. This challenge is effectively illustrated by the basic idea of "post-normal science", introduced by Funtowicz and

Ravetz (1993) to mark the "...passing of an age when the norm for effective scientific practice could be a process of puzzle-solving in ignorance of the wider methodological, societal, and ethical issues raised by the activity and its results" (p. 86).

The "post-normal" situation challenges the institutional division of labor that informs our everyday theories about how modern societies work and shape the interaction between experts and professionals. Instead of an imagined division of labor between different institutions and domains, we now have problems that force actors and the seemingly differentiated "logics" of separate domains to overlap. Problems present themselves as *risks* rather than *obstacles*, and we have arrived in a situation where "... facts are uncertain, values in dispute, stakes high and decisions urgent" (Funtowicz and Ravetz 1993: p. 86). Consequently, there is increasing pressure for researchers to move from the role of neutral expert to advocate for particular solutions. The role of socially responsible scientists (also called "concerned scientists") is not a new one, of course. Neat institutional boundaries have always partly been a modern illusion that has covered up real hybridity and boundary activity (Latour 1993, 2013). In the case of climate change, these tensions and boundaries have become exceptionally intense and important, partly because climate change has been a problem predominantly identified in the field of (natural) science (and less through other institutions or everyday experience). As the stakes for climate policy have increased, this has created added political pressure for scientists and the platforms where policy is crafted. Indeed, as an institution the IPCC itself functions as a hybrid platform where interaction between science and governments is built into its mandate. As this hybrid platform of knowledge production becomes the object of media reporting and public debate, there is a heightened sense that the negotiations that went into crafting this knowledge will be reopened in the public domain. In the recent history of climate politics, the 2009 case of hacked and leaked emails is an extreme example of this boundary leaking (Pearce 2010; Mann 2012).

Finally, climate change is a complex problem because of the way it exposes large-scale inequalities of the world in which we live. Beck (2010) underlines this insightfully stating that "...if we want to locate climate change at the heart of sociology *and* politics, we have to link it internally to the power and conflict dynamics of social inequalities" (p. 267). He sees social inequalities and climate change as "two sides of the same coin" (Beck 2010: p. 267), and as we begin to witness the rising risks and hazards of ongoing climate change, these issues will become ever more pronounced. The impacts, risks and levels of preparedness for forthcoming consequences are so unequally distributed, it is clear that climate change means very different things to different people. For some, climate change means a consumer choice of what kind of car would be "proper". For others, it can mean change of livelihood or the necessity of migration. The varieties of complexity discussed here point directly to the role of media and communication and pose difficult but inspiring questions: how to negotiate global knowledge with local sensibilities and the "spectral" demand? How to take into account the diversity of local political cultures and institutional arrangements without which necessary global agreements will be neither effective nor legitimate? How to imagine and develop new practices of cooperation and interaction between different stakeholders so that they retain their critical autonomy (of science, of journalism)? How to reimagine global professionalism—of scientists, journalists and so on—in a climate risk world?

Political Imagination: Planning, Challenging, Deliberating

Issues of scale and complexity play out in the many ways that climate change problematizes our thinking about *politics*. Some thinkers see climate change as a problem that underscores the need to overcome traditional frames of political thought. Giddens, for instance (2009: pp. 91–94), calls for a "post-political" consensus, combined with a return to a stronger role for the state in framing long-term political directions and facilitating change. He urges nation-states to start managing climate change as part of a more diverse set of risks and to promote more political and economic convergence. He wants to see states acting simultaneously as interested actors (because stakes are high) and uninterested facilitators (remaining neutral to particular solutions). A particularly interesting aspect of Giddens' argument is his call for a "return to planning" (Giddens 2009: p. 94). In this blueprint, a new kind of government planning is counter-balanced by the role of NGOs, businesses and other kinds of policy entrepreneurship. As part of this arrangement, Giddens also wants to partly de-politicize the issue of climate change arguing "...the issue is so important and all-encompassing that the usual party conflicts should be suspended or muted" (2009: p. 114).

Calling for such a post-political "concordat" illustrates how climate change provokes re-evaluations of what politics is good for, putting "democracy on hold" for a moment (Lovelock 2010; Randers 2015).

A provocative example that develops this line of thought is a recent essay by Oreskes and Conway (2014) which imagines global warming leading to the demise of "Western Civilization". Their fictional history of the "Second People's Republic of China" is set in the year 2393. It predicts that China, partly due to its undemocratic governance structure, is best equipped to adapt to rapid demands for policy change. In contrast, "Western Civilization", despite its public commitment to "knowledge", "science" and "evidence-based policies", was unable to gear into action quickly or effectively enough (see also Dryzek et al. 2013: pp. 94–101).

More or less explicitly authoritarian climate governance is not, however, the only imaginable solution. Some thinkers take an alternative path urging us to politicize climate change more thoroughly. In such views, bracketing partisan political energy from climate politics would not only be wrong but also fatal. This is Klein's (2014) argument for changing "everything". In more theoretical language, Mouffe's (2005, 2013) arguments against a "consensual" politics is that such visions always exclude important voices from social debate and cause new problems in place of those they claim to solve. In the context of climate change, a "post-political", state-sponsored "concordat", then, would necessarily exclude some actors, limit political motivation and generate antagonism toward the political efforts among the excluded. Taken on the scale of global policy formation, this political tension is of course clearly manifest in critical debates about the post-colonial legacy that also extends to climate issues. While perhaps in national arenas states can partly appear in the Giddensian role of facilitators, the Post-Westphalian power realities of global climate politics position them as self-interested actors. Thus, despite its merits of raising consciousness and building an essential sense of urgency, the idea of "solving" climate change with a globally binding transnational agreement, or with a political consensus (once agreed and then acted upon), is politically very problematic. Hulme (2014), for instance, argues for pluralism (in line with Mouffe) and propagates a pragmatist approach:

> …a world of more than 7 billion people cannot move together. Such a world will not agree on a single thermostat setting. The corollary of pluralism is philosophical and political pragmatism (…) Pragmatism is thus content to recognize and name problems like climate change as being super-wicked in character: non-definable and not solvable. Instead of using science and technology to "fix" wicked problems, pragmatism is content to pursue multiple and clumsy solutions to regularly reframed problems in order to achieve merely incremental gains. (Hulme 2014: pp. 137–138)

Illustrating a similar tendency from a different theoretical background, Ostrom (2009) has pointed out how "single policies" adopted at a global scale are "...unlikely to generate sufficient trust among citizens or firms so that collective action can take place in a comprehensive and transparent manner that will effectively reduce global warming". Instead, she argues for a "poly-centric approach" with active oversight of local, regional and national stakeholders:

> A polycentric approach has the main advantage of encouraging experimental efforts at multiple levels, leading to the development of methods for assessing the benefits and costs of particular strategies adopted in one type of ecosystem and compared to results obtained in other ecosystems. This problem [climate change], and having others also take responsibility, can be more effectively undertaken in small- to medium-scale governance units that are linked together through information networks and monitoring at all levels. (Ostrom 2009: p. 1)

The pragmatist and polycentric views offer crucial insights to climate politics. Theoretically, however, there are also deeper questions than asking whether we should put party politics on hold or at which levels and scales politicization and policy-making should take place. One such discussion emerges around the politics of epistemology. It is evident that pluralist, constructionist and pragmatist views have been extremely valuable in opening critical space for diversifying questions concerning power, history and justice, and they hold important potential for doing so in relation to climate politics. Yet, such tendencies can also run parallel and (however unwittingly) resonate with a tendency to deny the "reality" of climate change, or to blame scientists for constructing conspiracies. This is not to say that organized skepticism or climate denial was caused by academic theory, constructivist epistemologies or pluralist views of politics. Denialist movements and networks clearly emerge with massive systematic political and economic support from particular stakeholders, like the fossil-fuel lobby (Oreskes and Conway 2010; Dryzek et al. 2013: pp. 20–37), rather than from political or social theory. Their persistent grasp of parts of public opinion is linked to classic characteristics of individual psychology and its reactions to situations of dissonance between information and existing beliefs and behavior. Still, the more we have come to accept constructivist epistemologies in public life, the more difficult it has become to make scientifically valid authoritative claims about reality. In climate politics, *some kind* of shared knowledge about reality seems to be an essential ingredient if we are to begin coming to terms with what must be done. In this recognition of the shared realities—at least an agreement on the urgency for action—the IPCC has been an essential global organ, in particular toward the run up to the 2015 UNFCC Paris Conference of the Parties (COP21).

A third way of imagining climate politics can be based on the notion of *deliberation*. Dryzek (2013, 2015), for instance, has tried to articulate a more subtle road between constructivism and realism, arguing that effective climate action does not depend on reaching an ultimate consensus, nor that deliberative processes necessarily water down political energy and exclude relevant experience. Instead, he claims that well-facilitated deliberative processes can produce workable agreements between different viewpoints that allow for some crucial action to occur—without demanding that political pluralism (different notions of justice, for instance) be sacrificed.

Even a short discussion of the politics of climate debates reveals how the idea of the Anthropocene puts our notions of politics into new light. The fact that human activity has become a major, systemic driver of global natural mechanisms means that we are no longer merely "part of nature" but also increasingly a dominant factor of the nature in which we live. In this situation, it is essential not to remain stuck on earlier confrontations. This is especially important for studying the role of media and communication in politics and governance. It remains unclear how transnational, multi-sited, multi-scaled, networked and yet deliberatively functioning publics can come into being—and become powerful enough to inform politics effectively and legitimately. Yet, whatever the possible models might be, communication and journalism will be key factors for representing the world, and for situating various stakeholders within a deliberative space of mutual interpretation.

Climate change accentuates the contradictory ways in which modern societies think of politics inside nation-states, and in relation to global governance by extension. These strands of political theory formulate multiple challenges for journalism. Should journalism support the *general* authority of science but contribute to local criticism and "spectral" demands, to (further) politicizing science? Should journalism actively support "putting aside" big ontological-epistemological questions about climate change and instead concentrate on framing the public debate on *how* we should *react* to it (locally, nationally or globally)? How could journalism contribute—locally, nationally or globally—to a more deliberative relationship between institutions and stakeholders, including citizens, and to a participatory framework that would not ask questions of *effect* but of *participation*? These are just some examples of the questions this study tackles. Our main target in this book is to take the launch of the IPCC

AR5 and the work of this historically unique science-policy institution as a starting point. By following journalistic reactions and interpretations to this body of global knowledge, we want to describe, analyze and critically evaluate the performance of journalism *as a key part of a transnational communication infrastructure.*

Global climate politics has—after the Paris 2015 agreement—entered a phase of urgency with new kinds of governance challenges. In future mitigation efforts, a core mechanism of the new global agreement is built on accepting individual and differentiated contributions, which puts a new emphasis on *transparency* and *mutual monitoring.* In developing adaptation capacities and models, reaching toward a basic sense of fairness and *justice* is a necessary element of effective outcomes. Building a more nuanced and relevant roster of local policy models and solutions, *recognition* of all stakeholders as part of constructing the solutions is vital. All these broad preconditions of successful global climate politics are embedded in questions of communication.

CLIMATE CHANGE, MEDIA AND JOURNALISM

As climate change has become a recognized, agenda-setting issue for national and global policy, media and journalism research has begun to pay closer attention to problems of climate communication. Critical reviews of earlier media research problems have usefully summarized these challenges. Researchers have, for instance, pointed out how specific difficulties of climate communication relate to the nature of the problem itself (e.g. Moser 2010). Some have underlined the need to understand the whole "circuit" of climate communication politics (e.g. Carvalho and Burgess 2005; Berglez and Olausson 2014), or the media's role in the broader "cultural politics" of climate change (Boykoff 2011). Others call for a contextualization of climate and media with the political economy of global media and emerging strategic communication and lobbying practices (Anderson 2009). There have also been several invitations for more comparative research (Painter 2010) that looks beyond the Western print media (Schäfer and Schlichting 2014; Berglez and Olausson 2014; Painter 2010). Berglez and Olausson (2014) also emphasize the need for more interdisciplinary work on understanding climate media, urging researchers to work to reduce the theory–practice divide in media research.

This book offers a new contribution to this cumulating evidence about the key role of media and journalism in the politics of climate change.

We will engage with more specific arguments in this literature along the way. At the outset, however, we wish to underline several choices that shape the specific contribution of this study.

Global Geopolitical Reach

Because of its global scale and political complexity, climate change demands media research that covers a wide range of different conditions and political contexts. The specific aim of the *MediaClimate* network has been to cast a wide net over the world with a *transnational approach* that extends beyond Western nations. Our work draws from a long-term cooperation with scholars in more than 22 countries, covering all continents.

For a comparative project the diversity of contexts poses many challenges. These range from issues related to relevant sampling and coherent execution of comparative analysis to attempts to develop neat explanations of findings. For a comparative endeavor, the mere diversity of languages, journalistic cultures, and climate political contexts poses some limitations. Therefore, the strictly comparative results must be interpreted with caution. The core content analysis (see below, and Sidebar 1.1) that brings all the countries together aims at a rudimentary overview of the performance of mainstream journalism in 22 countries. Through a focus on mainstream press coverage of the IPCC AR5, we try to capture a snapshot of the *transnational communication infrastructure* in which global knowledge about climate change is circulated. In particular, we highlight and critically discuss the role of journalism in this system. We use our content analysis of this coverage as a *starting point* that helps us to formulate specific thematic questions. Through these questions, we draw on the expertise of network members working in particular contexts. The ultimate goal of the study, then, is to open up, elaborate and highlight particular moments and specific dynamics inside the climate communication infrastructure.

IPCC AR5 and the Dynamics of Global Media Events

The study at hand focuses on a particular, critical moment of time in the changing global landscape of climate politics. Focusing on the yearlong "event" of the AR5 launch, we build a specific case study on the role of global, scientific knowledge in journalism. The AR5 process itself was an unforeseen accumulation of scientific evidence and national political interests. This made the publication during 2013–2014 a unique "scientific event". With all its built-in compromises and political language, the publications of the AR5 reports constructed a highly loaded, imaginary common starting point for mapping the different journalistic reactions to it.

We look at the IPCC AR5 launch year, then, as a particular kind of global media event, a moment of anticipated, high public attention. This was critical moment both within the long process of the AR5 cycle itself, and in the broader context of global climate politics. As the actors involved in the AR5 and its communication processes knew, the previous high tide of global climate attention (climaxing in 2009) proved problematic for climate science, for the IPCC and for the global process climate governance. The "Climategate" email leaks and disputes surrounding mistakes in the previous IPCC AR4 were a hard lesson about the volatility of the forces that are in play in the current global media environment. Public disappointment after the failure of the Copenhagen climate negotiations—in relation to the high expectations and hopes raised in global media—was still rather fresh in memory. Hence, the stakes at the AR5 launch were particularly high. The publication promised to be another test encounter with climate skeptics and other interest groups working against coordinated intergovernmental climate action. It was also a crucial moment for building the knowledge base on the road to the Paris climate summit (COP21) that was still ahead.

As such, the AR5 launch event provides an example of a particular kind of contemporary *global media event*. This notion, originally coined by Dayan and Katz (1992), has in recent decades helped media researchers to focus on different kinds of exceptional moments of concentrated public attention. Initially, the key focus was on nationally encompassing moments whereby mass media were regarded as instrumental in connecting large populations to the imagined, "centre" of society (cf. Couldry 2003). In recent years, however, the debate has been more focused on the transnational, disruptive and unpredictable dynamics of such moments (Hepp et al. 2010). This combined with the fact that attention grasping events in the current globalized environment cannot be easily—even nationally—managed by individual actors as effectively nor as easily as in the past. Today media events often become moments of breakdown rather than celebration of social order (Katz and Liebes 2010). Despite this complexity—or perhaps because of it—the notion of media events remains useful. In understanding global communication, it underscores the way in which transnational communication infrastructures (see above) are activated in particular, always partially contingent conjunctures. Understanding the dynamics of the event is crucial for making sense of the communication environment in which global climate science and climate politics is embedded.

MediaClimate network is the international research collective behind this study. Earlier we examined global climate journalism by comparing coverage of the COP summits, particularly in Copenhagen (2009) and Durban (2011) (Eide et al. 2010; Eide and Kunelius 2012).[1] As highly constructed, organized and orchestrated political episodes, such "mega-events" provide particularly intense and interesting moments to analyze and compare the performance of journalism (see also Painter 2010; Wessler 2012; Lück et al. 2015). In this book, we use the launching of the IPCC AR5 (2013–2014) to focus on a different kind of moment in global climate communication. Focusing on the AR5, the study looks into how journalism mediates and translates the "global knowledge" of the IPCC into the public domain. We look at how journalism reacted, globally, to the synthesis of scientific-political knowledge produced by the IPCC. The "extended event" of the AR5 launch (during 2013 and 2014) provided a chance for us to capture some of the characteristic similarities and differences in how journalism makes sense of climate change in general and climate science in particular. It was a pre-scheduled, anticipated moment of global attention that drew together massive knowledge resources, attracted lots of interest in shaping public interpretations of the results, and created possibilities for journalists to collaborate. At the same time, it put an immense amount of pressure on the performance of journalism.

Mainstream Print Bias and the Notion of the "Public"

At the core of this research project is a global sample of *mainstream newspaper* coverage of the IPCC AR5 in 22 countries. To study the mainstream press in today's media environment is obviously to claim that it still matters. As critical reviews of media research point out, this is no longer self-evidently true (presuming it ever was). Local media infrastructures differ throughout the world, and the role of newspapers as dominant forms of public communication has never been universal. Certainly, with recent digital transformations, this landscape is even more complex, casting doubt on the centrality of the press and mainstream journalism in general. Our empirical starting point does, however, use the mainstream press to capture *reactions* in the brand of *professional journalism* that works closely with elites and networks of key stakeholders in climate politics.

[1] For a more exhaustive list of publications by the *MediaClimate* network, see mediaclimate.net

We approach this coverage, then, not as evidence of the *transmission* of knowledge and opinions from the IPCC to an assumed general audience. Rather, we work on the assumption that professional mainstream journalism offers relevant evidence of how the media, and journalists in particular, build a visible, social and political *space of interpretation* where the discussion of climate change takes place. In this space, we take interest in the different stakeholders and actors who voice their interpretations of climate politics.

In order to map this space globally, two newspapers were selected from each of the 22 participating countries (Table 1.1) with the aim of capturing a fair sample of mainstream, "quality" journalism in each location. Two newspapers, obviously, do not always encompass the entire political spectrum in each national media sphere, nor do they represent the diversity of journalism in the quality-popular dimension. Thus, we only capture a slice of the mediated space of climate politics that is by no means exhaustive. One need only think of all the debates about climate in social media (e.g. Steskal and Diakopoulos 2014; Pearce et al. 2014) to realize that the sample, at best, presents a strategic glimpse of contemporary climate communication. However, we take it to offer important information about politically relevant and consequential aspects of media discourse, due to its proximity to political decision-making. While it does not show what people actually think, the mainstream press represent two kinds of "public opinion" in one space: (1) the public opinions of powerful elites and (2) the imagined opinions "people" would presumably have if they had enough time to inform themselves of important issues (see Nerone 2015: pp. 1–9). Hence, we take the print bias seriously but also believe that mainstream journalism still offers a relevant object of study. We know—from a long tradition of journalism research—that the diverse voices and viewpoints in news journalism tend to "index" the opinions and views of elite power stakeholders in national public spheres (Bennett 1991). Thus mainstream journalism serves as one way of tracking balances of power in politics and climate governance, and the "legitimate sphere of controversy" (Hallin 1986) on a given issue in a given national context. By mapping mainstream journalism, then, we hope to capture something more general about the state of consequential public debates in different corners of the world and the relationship of journalists to other power agents in climate politics. Hence, while the space of interpretation

Table 1.1 Overview of the total sample (countries, media, number of stories)

Higher coverage			*Lower coverage*		
Australia	Daily Telegraph	85	South Africa	Business Day	24
	Sydney Morning Herald	190		Cape Times	12
USA[a]	New York Times	120	Bangladesh	Daily Star	21
	Washington Post	71		Prothom-Alo	12
Japan	Asahi Shimbun	82	Russia[b]	Kommersant	14
	Yomiuri Shimbun	89		Rossiiskaia Gazeta	16
Brazil	O Estado de São Paulo	108	Chile	El Mercurio	3
	O Globo	60		La Tercera	23
Finland	Aamulehti	48	Uganda	Monitor	12
	Helsingin Sanomat	86		New Vision	10
Canada	Globe and Mail	47	China	China Youth Daily	7
	Toronto Star	49		People's Daily	13
UK[c]	Guardian	50	Indonesia	Kompas	10
	Telegraph	38		Koran Tempo	4
France	L'humanité	39	Ethiopia	Daily Monitor	5
	Le Monde	46		Ethiopian Herald	8
Norway	Aftenposten	35	Israel	Haaretz	6
	Dagsavisen	44		Yedioth Aharonoth	1
Sweden	Dagens Nyheter	43	Poland	Gazeta Wyborcza	5
	Svenska Dagbladet	28		Rzeczpospolita	2
Germany	Die Tageszeitung	26	Egypt	Daily News Egypt	3
	Süddeutsche Zeitung	18		Cairo Post	2

[a]The US numbers for *The New York Times* include 21 stories from the newspaper blogs section and 6 stories from its affiliates *The International Herald Tribune* and *The International New York Times*. The *Washington Post* number includes 17 online stories.
[b]The largest part of Russian coverage, 11 stories in *Kommersant* and 9 in *Rossiiskaia Gazeta*, were only published online.
[c]*The Guardian* sample includes 6 stories published in *The Observer* which is considered part of the newspaper.

we capture is limited, it arguably maps important contours of the public debate—and these structures of meaning and public action have effects beyond the confines of mainstream print media.[2]

We approach the AR5 (2013–2014) by focusing on coverage of its four sub-reports. The main sample of newspaper coverage was collected around the publication of each report, beginning a week prior to the scheduled

[2] In many European countries, newspapers continue to be read by a large proportion of the population. In developing countries, where illiteracy and poverty are widespread, they cater (even more) to elites. There are several reasons why studying newspapers remains our priority. One is the availability of the material. Another is that newspapers, even if poorly

launch and for three weeks after each launch. In the 22 countries this amounted to 1615 stories. This full sample consists of four types of stories, categorized by their relevance to the central event, the IPCC report launches (Table 1.2). Thus, when we elaborate on the results of the media sample, we refer to four distinct sub-samples:

1. The IPCC-core sample (N = 551) includes all stories where the IPCC AR5 was the main topic, usually already mentioned or strongly referred to in the headline.
2. The AR5-sample (N = 780) includes stories focusing more generally on climate change as well as those that refer to IPCC AR5 explicitly, even if only briefly.
3. The general climate sample (N = 1254) includes stories focusing on climate change that do not refer to the AR5.
4. Total sample (N = 1615) all stories, included those that only refer to climate change and/or the IPCC but are mainly focused on another topic or event.

The Space of Interpretation: Attention and Access

The project analyzes this mainstream journalism content by applying a set of quantitative and qualitative perspectives. We approach the space of interpretation constructed by mainstream journalism first with an empirical strategy that can be presented on two levels.

The first level of analysis looks at the reaction of journalism through the notion *attention*. The role of media and, in particular, the role of *mainstream* media are closely linked to its power to manage the imagined focus or topic of public attention in a given moment. Earlier studies of climate change media show that public attention has oscillated considerably (e.g. Boykoff 2011: pp. 20–29; Schmidt et al. 2013; Schäfer et al. 2014; Eide and Kunelius 2012), reflecting both what happens in

---

distributed among the population at large, still play an important role for political and cultural elites—and their debates—in any country. In some countries it was difficult to restrict the study to print newspapers due to the dominant role played by online versions. Thus, some of our results are higher than they would have been if only print items were considered. In addition, one member of the research network has done separate (limited to fewer countries and a shorter time span) research on television coverage of the four IPCC reports (Painter 2014, 2015a, c).

Table 1.2 Number of stories in the four sub-samples

	Core	AR5	General climate	Total
	(1)	(2)	(3)	(4)
Australia	44	65	183	275
Bangladesh	8	18	30	33
Brazil	87	105	145	168
Canada	22	41	68	96
Chile	10	20	25	26
China	4	5	12	20
Egypt	3	4	5	5
Ethiopia	1	5	13	13
Finland	48	68	111	134
France	40	52	73	85
Germany	23	31	39	44
Indonesia	12	14	14	14
Israel	4	4	7	7
Japan	50	63	94	171
Norway	36	58	73	79
Poland	3	7	7	7
Russia	10	11	24	30
South Africa	20	26	35	36
Sweden	43	57	65	71
Uganda	1	1	5	22
UK	59	79	86	88
USA	23	46	140	191
Total	551	780	1254	1615

the field of climate politics (e.g. global summits) and what other issues compete for limited media attention. By recognizing issues as noteworthy, the mainstream media creates the primary condition of publicity: the attention of the "public" to a problem, question, or topic. It is this attention that builds the moment and momentum of the mediated public sphere, raising the public and political stakes for politicians and other stakeholders who engage in the public debate. In other words, it is this imagined public attention that invests the public sphere with its political importance. Our comparative content analysis of the global sample assigns attention a score that combines findings about the amount of stories and their size (Sidebar 1.1).

Sidebar 1.1
MediaClimate IPCC AR5 content analysis

The analysis focused on two mainstream newspapers in 22 countries (Table 1.1) during the four data collection periods around each Working Group (WG) report:

- 24 September–15 October 2013 (IPCC WGI: The Physical Science Basis)
- 29 March–12 April 2014 (IPCC WGII: Impacts, Adaptation, and Vulnerability)
- 13 April–1 May 2014 (IPCC WGIII: Mitigation of Climate Change)
- 30 October–20 November 2014 (IPCC AR5 Synthesis Report)

All stories mentioning "IPCC", "climate change" or "global warming" were collected for the *total sample* (1615 stories). This sample includes all stories somehow related to climate change, even if by brief mention, as well as any material unrelated to the AR5. By excluding stories that only briefly mention climate change or IPCC, and those that focus on something else, we limited this to the *general climate sample* (1254 stories). Inside this general sample, we reduced the data further to stories on climate change that mention the IPCC, the *AR5-sample* (780 stories). *We mostly focus on this sample.* Inside the AR5 sample, the *core sample* (551 stories) contains stories with IPCC AR5 as the main topic. In some cases, the core sample is more relevant to our analysis. Comparison of these different samples suggests that AR5 stories accounted for 62 percent of the total media coverage during these periods.

In this book we use the findings from our content analysis in the following manner.

Attention score. Findings on the amount of *attention* devoted to AR5 in different countries are measured by multiplying the number of stories and the average story size during specific WG periods. Story size was coded as 1, 2 or 3 (less than 300 words, 300–600 words or more than 600 words). The resulting attention score is a rough indicator of the amount of attention. This figure, however, does not relate the amount of IPCC AR5 coverage to the overall space available (the daily news hole). We use the attention score to

(continued)

Sidebar 1.1 (*continued*)

roughly group countries by media attention and the weight of IPCC knowledge in the national public (mainstream news) agenda. The attention figure is also used in order to track the dynamics between different WG reports.

Story genre. All stories were coded according to their journalistic genre. An initial code scheme differentiated of ten categories (news, reportage, interviews, profiles, background stories, graphs and fact boxes, external op-eds, letters to the editor, editorials, and staff columns). This coding was used to locate sub-samples for qualitative analysis. For quantitative purposes, we only used the collapsed categories and the dichotomy between *reporting* genres and *opinionated* genres (see Appendix). In different journalistic cultures across the globe, the borderline between stories where journalists are more visible varies a lot. In some countries, news stories allow journalists a more assertive voice. While this blurs distinctions between some reporting genres, the key opinionated genres can be identified more reliably.

Voice distribution. Coding of the voices aims to provide a quantitative measure of the access different kinds of stakeholders have to the public space opened up in the coverage. All stories were coded by identifying *individual people* (mentioned by name) and *quoted* (either directly or indirectly). While restricting the sample to named voices does leave out some anonymous sources, this strategy was chosen to provide maximum consistency in coding. The voices were coded according to the following main distinctions:

1. Sector of society was identified by the main categories (national political system, transnational political system, civil society, business, science and media) and by more elaborated distinctions under each main category (Table 1.3). Here, again, in the quantitative description of the material, we use mostly the main categories. For instance, President Barack Obama and a national member of parliament would be coded as *national political system*, UN or EU representative as *transnational political system*, local grassroots actors and Greenpeace representatives as *civil society* and so on. Business actors belonged to

(*continued*)

Sidebar 1.1 (*continued*)

business when representing a company and to civil society if speaking on behalf of a wider association.

2. Domestic-foreign distinction was identified by judging from the location of the newspaper. This aims at tracking the importance of domestic news sources and their network. Thus, *national* political actors were identified either as domestic of foreign. Barack Obama could therefore be coded either as a foreign or domestic actor, depending on where the news story was published.

3. Scientific actors were coded according to their specific area of scientific expertise. Unclear cases were coded under the category other. Here we mainly use the distinction between natural scientists and other sciences in order to reflect on the dominance of natural science in the coverage. In order to track the IPCC's role in the coverage, we also made note of those scientists who were *explicitly* connected (in the text) to the IPCC and of those who were not. This allowed us to reflect on the role of local IPCC networks in efforts to interpretation the reports.

4. Gender of the news voices was also coded. Having local coders assured that this, mostly name-based, coding was achieved.

Reliability challenges. Reliability of the coding between 22 countries was rehearsed with the research team at two international workshops (Paris, April 2014, and Tampere, April 2015) where problematic cases and details were discussed. The consistency of the coding through the 22 native language coders remained a challenge throughout. In order to produce a coherence coding result, all voice coding decisions were double-checked by comparing the *names* and *institutional* affiliations (recorded in the coding scheme) of the voices against the coding decisions made by different teams. Thus, our team at the University of Tampere verified and checked for consistency in the AR5 sample (780 stories, 1332 voices). Because of potential reliability issues, we avoid using the data to analyze complex patterns of relationships between

(*continued*)

Sidebar 1.1 (*continued*)

countries. Instead, we look at the main voice categories to provide broad discussions of the country profiles (Chap. 4).

Headline highlights: As an additional aspect of *attention* analysis, headlines from the core sample (551 stories) were coded using English language translations provided by members of research network. These are only used to analyze global press reactions to the different WG reports. The headline coding is elaborated in Chap. 3 where it was applied.

Themes. In some countries, a more elaborated coding scheme was used to elaborate on how the coverage emphasized particular themes (disaster, uncertainty, skepticism, risk, justice, opportunity from action and opportunity from inaction). This coding is only used in Chap. 5 where the coding details are explained.

The second level of analysis concentrates on *access*. This refers to the way in which journalism invites and pressures other actors to appear in the social space opened by the media attention. What makes the public space political is that it is inhabited by different stakeholders. In addition to being able to focus public attention then, the power of journalism is illustrated by its ability to manage the access and appearance of other actors. Public sphere theories often remind us of the twofold nature of access, or of being "in the public". On one hand this means being monitored, and on the other hand it means being engaged in the process of policy formation and being recognized as a relevant (powerful) actor. Although new infrastructures and conditions have made things more complex, this classic dualism of the social imaginaries in public spaces or spheres (e.g. Taylor 2004; Splichal 2009)—the public sphere as both a sphere of *monitoring* and a sphere of *engagement*—still presents a valid way of making sense of how the "global public sphere" works (Volkmer 2015). In order to analyze the access of news actors to the mainstream news flow, our study identified and coded all individuals who were quoted (either directly or indirectly) by name in the global sample. These explicitly cited "voices" do not, of course, serve as a direct indicator of the actor networks and co-productive relationships of local fields of climate politics that underlie journalism. However, looking at similarities and differences between countries, between different AR5 themes, and between the different categories of actors (politicians, scientists and civil society actors) that dominate (or do not) in this public space provides fruitful points of departure for gaining insights on the relationship between media, science and politics.

Table 1.3 offers an overview of the distribution of all the voices in the total and AR5 samples. It shows the strong role of scientists and points

Table 1.3 Distribution of voices (percentage within the AR5 (780 stories, 1332 voices) and total sample (1615 stories, 2725 voices)

Voice	AR5	Total
National political actors	**17.2**	**28.2**
Domestic	10.9	17.6
Foreign	5.6	9.9
Other	0.7	0.7
Transnational political actors	**11.4**	**7.6**
Regional institutions	2	1.4
UN	8.9	5.6
Other	0.6	0.6
Civil society	**16.9**	**17.7**
Domestic	10.7	11.4
Foreign	3.3	3.1
NGO	7.9	7.4
Think tank	2.1	2.2
Ordinary people	6.2	7.2
Other	0.8	1
Business	**3.2**	**5.3**
Domestic	1.9	3.2
Foreign	1.2	1.9
Other	0.2	0.2
Science	**47.8**	**35.5**
IPCC	24.5	12.3
Domestic	28.6	22.7
Foreign	19.2	12.8
Natural science and technology	31.3	22.2
Social science and humanities	3.1	3.2
Economics	4.4	4
Other	9	6.2
Media	**3.1**	**4.1**
Other	**0.4**	**1.6**

Note: Percentages refer to the share of voices in the whole sample to the overwhelming dominance of natural science. It also points to the importance of national political actors and civil society voices. This global distribution is, of course, an interesting indicator, however abstract, of the overall performance of mainstream journalism. It offers a partial representation of the infrastructure of global climate media, and its structural biases. The structural characteristic of this journalistically created space is a key object of this study.

We approach the global structural features of climate journalism in several chapters. Chapter 4 provides an overview of the differences between the 22 countries in the sample. It also considers a wide range of background factors that can have a bearing on both the amount of attention and the diversity of voices (access) in national spheres of publicity. Drawing from theorizations on how media *domesticate* global news and events, the chapter distinguishes four preliminary ideal types, or *mediated civic epistemologies*, that characterize certain groups of countries. Chapter 5 takes the analysis to a more detailed, thematic level by comparing how particular themes (ranging from disaster to risk, skepticism and global justice) were articulated in the AR5 coverage. By comparing the performance of developed and developing countries, the chapter looks at the uneven emphasis of different themes and contributes to a growing body of literature on how the media *frames* climate change. Chapter 6 elaborates further on some key thematic and conceptual debates in global journalism by concentrating on a selection of opinionated stories from the AR5 sample. The chapter draws on theories of *justice* and *solidarity* to critically evaluate the way the mainstream press articulated an inadequately developed framework of global solidarity. Chapter 7, in turn, groups together countries belonging to the emerging economies which have played a crucial role in global climate politics in the post-Kyoto period. It illuminates the search for compromise between the right to develop and the global urgency of the climate risk. Comparison of coverage in these countries shows that simple country groupings, according to broad global climate politics, do not offer plausible explanations for coverage. Instead, there are considerable differences that need to be understood against local political struggles and stakes.

Moments in the Communication Process

While the core structure of this study is anchored around the content analysis and journalistic material it captures, the book casts a wider, more nuanced net over the event dynamics and the networks of actors involved. We look at climate communication from the point of view of IPCC scientists (Chap. 2) and key journalists (Chap. 12) by conducting interview research. In both cases, these key actors in the IPCC AR5 communication process are asked to reflect upon their mutual relationship, based on a certain amount of tension between them, but also on shared interests. In addition, we broaden the context of interpretation by discussing two kinds of networks around the core journalism contents: policy networks that interact with local journalism (Chap. 8) and digital networks that re-circulate mainstream journalism and function as an alternative infrastructure of links and connections (Chap. 9). Focusing on these additional moments in the circuit of communication does not, of course, grasp the complexity of the IPCC AR5 release event. However, they do help to situate the performance of journalism, and raise new questions that will remain on the agenda of further research.

Professional Challenges

Since the study focuses on journalism in particular, we also highlight some fundamental issues related to *professionalism*. Chapter 10 analyzes coverage in several developing countries and looks at IPCC coverage by focusing on a particular strand in the professionalism debate: namely the idea of *development journalism*. The chapter identifies some of the strengths and weaknesses of that notion when it comes to climate science communication. Chapter 11 focuses on the professional quality of the coverage by identifying what we consider to be particularly insightful examples of "good practices" in our mainstream sample. These examples demonstrate the creative power of journalists and their professional imagination. They also show that exceptionally good coverage can spring up in circumstances where structural factors—such as time and newsroom resources—remain quite limited. Chapter 12 draws on interviews conducted with some of the key journalists who covered the AR5 globally. It takes the issue of climate change—the scale and depth of the problem—as a potential source for locating new kinds of professional practices and values.

This book aims at an empirically and theoretically rich reading of what took place in the transnational space of interpretation. It does so through an analysis that is anchored by the IPCC AR5 event and its reception in mainstream journalism. Individual chapters in the book play different roles in this overall framework, combining several different levels of argumentation and analysis. Many draw on the results of the general content analysis (Sidebar 1.1) but most bring in qualitative evidence and draw on additional materials. They also bring diverse groups of authors from different corners of the world into the analysis. We conclude with a brief discussion of some of the lessons and implications this work offers for the IPCC and its communication efforts, for journalism, and for media and journalism research. These conclusions are framed by the call of several experts who argue that, in the future, the IPCC must become more *agile*, *interdisciplinary*, *inclusive* and *comprehensible*. With some slight modifications, these demands would serve as qualities for better climate journalism as well. This book is an attempt to map out some of the opportunities and obstacles associated with achieving these ends.

BIBLIOGRAPHY

Anderson, A. (2009). Media, politics and climate change: Towards a new research agenda. *Sociology Compass, 3*, 166–182.

Beck, U. (2010). Climate for change or how to create a green modernity? *Theory Culture & Society, 27*(2–3), 254–266.

Berglez, P., & Olausson, U. (2014). The post-political condition of climate change: An ideology approach. *Capitalism Nature Socialism, 25*(1), 54–71.

Boykoff, M. T. (2011). *Who speaks for climate? Making sense of media reporting on climate change*. Cambridge: Cambridge University Press.

Carvalho, A., & Burgess, J. (2005). Cultural circuits of climate change in U.K. broadsheet newspapers, 1985–2003. *Risk Analysis, 25*(6), 1457–1469.

Dietz, S., & Stern, N. (2014). Endogenous growth, convexity of damages and climate risk: How Nordhaus' framework supports deep cuts in carbon emissions. *Working Paper No. 180*, Centre for Climate Change Economics and Policy. Retrieved from http://www.lse.ac.uk/GranthamInstitute/publication/endogenous-growth-convexity-of-damages-and-climate-risk-how-nordhaus-framework-supports-deep-cuts-in-carbon-emission/

Dryzek, J. S. (2013). The deliberative democrat's idea of justice. *European Journal of Political Theory, 12*(4), 329–346.

Dryzek, J. S. (2015). Reason and rhetoric in climate communication. *Environmental Politics, 24*(1), 1–16.

Dryzek, J. S., Norgaard, R. B., & Schlosberg, D. (Eds.). (2011). *The Oxford handbook of climate change and society*. Oxford: Oxford University Press.

Eide, E., Kunelius, R., & Kumpu, V. (Eds.). (2010). *Global climate, local journalisms: A transnational study of how media make sense of climate summits*. Bochum: Projekt Verlag.

Fernández-Armesto, F. (2011). *The world: A history*. Boston: Pearson.

Fraser, N. (2007). Transnational public sphere: Transnationalizing the public sphere: On the legitimacy and efficacy of public opinion in a Post-Westphalian World. *Theory, Culture & Society, 24*(4), 7–30.

Fraser, N. (2014). Transnationalizing the public sphere: On the legitimacy and efficiency of public opinion in a post-Westphalian world. In K. Nash (Ed.), *Transnationalizing the public sphere* (pp. 8–42). Cambridge: Polity.

Funtowicz, S., & Ravetz, J. (1993). Science for the post-normal age. *Futures, 25*(7), 739–755.

Giddens, A. (2009). *The politics of climate change*. Cambridge: Polity.

Habermas, J. (2001). The postnational constellation and the future of democracy. In M. Pensky (Ed.), *The postnational constellation political essays* (pp. 58–112). Cambridge: Polity Press.

Hulme, M. (2010). Cosmopolitan climates: Hybridity, foresight and meaning. *Theory, Culture and Society, 27*(2), 267–276.

Hulme, M. (2014). *Can science fix climate change? A case against climate engineering*. Cambridge: Polity Press.

IPCC. (2014, March 31). *IPCC report: A changing climate creates pervasive risks but opportunities exist for effective responses*. Geneva, Switzerland: IPCC. Retrieved from http://www.ipcc.ch/pdf/ar5/pr_wg2/140330_pr_wgII_spm_en.pdf

Jasanoff, S. (2011). Cosmopolitan knowledge: Climate science and global civic epistemology. In J. S. Dryzek, R. B. Nordgaard, & D. Schlosberg (Eds.), *The oxford handbook of climate change and society* (pp. 129–143). Oxford: Oxford University Press.

Klein, N. (2014). *This changes everything: Capitalism vs. the climate*. London: Penguin Books.

Latour, B. (1993). *We have never been modern*. Cambridge, MA: Harvard University Press.

Latour, B. (2013). *An inquiry into modes of existence: An anthropology of the moderns*. Cambridge, MA: Harvard University Press.

Lorenzoni, I., & Hulme, M. (2009). Believing is seeing: Laypeople's views of future socio-economic and climate change in England and Italy. *Public Understanding of Science, 18*, 383–400.

Lück J., Wozniak, A. & Wessler, H. (2015). Networks of coproduction: How journalists and environmental NGOs create common interpretations of the UN climate change conferences. Unpublished manuscript.

Mann, M. (2013). *The sources of social power: Volume 4, globalizations, 1945–2011*. Cambridge: Cambridge University Press.

Mann, M. E. (2012). *The hockey stick and the climate wars: Dispatches from the front lines*. New York: Columbia University Press.

McNeil, R. J., & McNeil, W. H. (2003). *The human web: A bird's eye view of world history*. New York: W.W. Norton.

Moser, S. C. (2010). Communicating climate change: History, challenges, process and future directions. *WIREs Climate Change, 1*(1), 31–53.

Mouffe, C. (2005). *On the political*. London: Verso.

Mouffe, C. (2013). *Agonistics: Thinking the world politically*. London: Verso.

Nerone, J. (2015). *The media and public life: A history*. Cambridge: Polity.

Nordhaus, W. D. (2008). *A question of balance: Weighing the options on global warming policies.* New Haven: Yale University Press.

Oreskes, N., & Conway, E. (2010). *Merchants of doubt: How a handful of scientists obscured the truth on issues from tobacco smoke to global warming.* New York, NY: Bloomsbury Press.

Oreskes, N., & Conway, E. M. (2014). *The collapse of western civilization: A view from the future.* New York: Columbia University Press.

Ostrom, E. (2009). A polycentric approach for coping with climate change. *Policy Research Working Paper 5095.* Background paper to the 2010 World Development Report. Released by The World Bank, Development Economics, Office of the Senior Vice President and Chief Economist, October 2009. Retrieved from http://citeseerx.ist.psu.edu/viewdoc/download?doi=10.1.1.1 77.7104&rep=rep1&type=pdf

Painter, J. (2010). *Summoned by science: Reporting climate change at Copenhagen and beyond.* Oxford, UK: Reuters Institute for the Study of Journalism.

Painter, J. (2014). *Disaster averted? Television coverage of the 2013/14 IPCC's climate change reports.* Oxford, UK: Reuters Institute for the Study of Journalism.

Painter, J. (2015a). Disaster, uncertainty, opportunity, or risk? Key messages from the television coverage of the IPCC's 2013/14 reports. *METODE Science Studies Journal, 85,* 73–79.

Painter, J. (2015c). Taking a bet on risk. *Nature Climate Change,* , 286–288.

Pearce, F. (2010). *The climate files: The battle for the truth about global warming.* London: Guardian Books.

Pearce, W., Holmberg, K., Hellsten, I., & Nerlich, B. (2014). Climate change on Twitter: Topics, communities and conversations about the 2013 IPCC Working Group 1 report. *PLoS ONE, 9*(4), 1–11.

Piketty, T. (2014). *Capital in the twenty-first century.* Cambridge, MA: The Belknap Press of Harvard University Press.

Randers, J. (2015, January 15). Demokratin har svårt att hantera klimathotet [Democracy is having problems managing with climate threat]. Released by *Extrakt*, published by The Swedish Research Council Formas. Retrieved from http://www.extrakt.se/debatt-opinion/demokratin-oformogen-att-hantera-klimathotet/

Sagoff, M. (2011). The poverty of climate economics. In J. S. Dryzek, R. B. Norgaard, & D. Schlosberg (Eds.), *The Oxford handbook of climate change and society* (pp. 55–67). Oxford: Oxford University Press.

Schäfer, M., Ivanova, A., & Schmidt, A. (2014). What drives media attention for climate change? Explaining issue attention in Australian, German and Indian print media from 1996 to 2010. *International Communication Gazette, 76*(2), 152–176.

Schäfer, M., & Schlichting, I. (2014). Media representations of climate change: A meta-analysis of the research field. *Environmental Communication, 8*(2), 142–160.

Schmidt, A., Ivanova, A., & Schäfer, M. S. (2013). Media attention for climate change around the world: A comparative analysis of newspaper coverage in 27 countries. *Global Environmental Change, 23*(5), 1233–1248.

Splichal, S. (2009). "New" media, "old" theories: Does the (national) public melt into the air of global governance? *European Journal of Communication, 24*(4), 391–405.

Stern, N. (2007). *The economics of climate change: The Stern review.* Cambridge: Cambridge University Press.

Taylor, C. (2004). *Modern social imaginaries.* Durham: Duke University Press.

Urry, J. (2010). *Climate change and society.* Cambridge: Polity Press.

Volkmer, I. (2015). *The global public sphere: Public communication in the age of reflective interdependence.* Cambridge: Polity Press.

Wessler, H. (2012). Identifying global public sphere moments. In J. F. Hovden (Ed.), *Hunting high and low. Skriftfest til Jostein Gripsrud pa 60-arsdagen* (pp. 437–455). Oslo: Scandinavian Academic Press.

CHAPTER 2

Scientists, Communication and the Space of Global Media Attention

Elisabeth Eide

> *Among us there is no one with professional communication skills, we are all researchers who try to express ourselves with clarity.*
>
> *Eystein Jansen (Interview, 13 May 2015)*

When the IPCC launched its Fourth Assessment Report (AR4) in 2007, climate change was high on media and political agendas (Boykoff et al. 2015), and global optimism surged. IPCC chair, Rajendra Pachauri, received the Nobel Peace Prize with the former US Vice President Al Gore, who had released his Oscar winning *An Inconvenient Truth* (2006). The UNFCCC Bali climate summit (COP13) ended optimistically in December 2007. A plan was in the making for the post-Kyoto era, and expectations for the Copenhagen summit (COP15) were on the rise. The AR4 concluded stronger than ever before: "Most of the observed increase in global average temperatures since the mid-twentieth century is

E. Eide (✉)
Department of Journalism and Media Studies, Oslo and Akershus University College of Applied Sciences, Oslo, Norway

© The Author(s) 2017
R. Kunelius et al. (eds.), *Media and Global Climate Knowledge*,
DOI 10.1057/978-1-137-52321-1_2

33

very likely due to the observed increase in anthropogenic greenhouse gas concentrations".[1] In 2013, the IPCC's Fifth Assessment Report (AR5) raised "very likely", indicating more than 90 percent scientific certainty, to "extremely likely" (95 percent certainty). AR5 WGI concludes: "This evidence for human influence has grown since AR4. It is *extremely likely* that human influence has been the dominant cause of the observed warming since the mid-twentieth century" (IPCC 2013: p. 15).

Simple logic suggests that raising certainty would increase the world's willingness to act. However, such a logic implies just as simple an idea of one-way communication, assuming that "...the provision of scientifically sound information can change public behavior and increase support for new policy measures" (Hulme 2009a: p. 217). The relationship between knowledge, public perception and our willingness to act is more complicated both psychologically and geopolitically (e.g. Capstick et al. 2015; Lee et al. 2015). Thus, the common sense model of knowledge flowing from science through journalism to an informed citizenry does not work (Secko et al. 2013). Rising scientific consensus on climate change does not guarantee a better-informed and increasingly engaged public (Austgulen and Stø 2013).

Despite the complexities between science, media and public awareness, how the IPCC communicates its results remains of great importance. As leading IPCC author Chris Field iterates, the IPCC occupies a unique position in climate change communication. The panel is in deep partnership with governments, and protecting this relationship is critical for governments and the IPCC. The assessment reports present a crucial "opportunity for climate science to speak directly to power" (Black 2015), creating a knowledge base upon which politicians can act. Mainstream journalism, in turn, provides an important catalyst for this science-policy link. It not only informs citizens but also *represents* the public, acting as a proxy for "...the way people would think if they would have enough time and knowledge" (Nerone 2015: p. 1).

This chapter focuses on the IPCC, its structure and its efforts to communicate the AR5 reports. It draws on science communication theories as well as interviews with several IPCC scientists mainly conducted during the UNESCO scientific conference "Our Common Future" in Paris, July 2015 (For a full list of the interviewees, see Sidebar 2.1). The

Sidebar 2.1
Interviews with IPCC scientists
In 2014 and 2015, we interviewed eight scientists and the IPCC communication director about their views on the IPCC process and its communication challenges. Most of the interviews took place on 7–10 July 2015 at a UNESCO conference on climate science in Paris. The interviews were semi-structured, all lasting between thirty minutes and two hours. The selection of respondents aimed at variety of voices, by region, gender and scientific discipline. The questions focused on how the IPCC planned its communication activities; on how they managed their position as "policy-relevant" but not "policy prescriptive"; and finally on how individual scientists viewed journalistic coverage of their science.

List of respondents:
IPCC communication director Jonathan Lynn (UK). Interviewed in Oslo, 20 September 2014 by EE and RK.

IPCC WGI author and co-chair Thomas Stocker (USA). Interviewed in Paris, 7 July 2015 by EE and RK.

IPCC WGI author Seita Emori (Japan). Interviewed in Paris, 9 July 2015 by RK.

IPCC WGI author Valérie Masson-Delmotte (France). Interviewed in Paris, 9 July 2015 by EE. Masson-Delmotte was recently elected co-chair of WGI for IPCC AR6.

IPCC WGII author Karen O'Brien (Norway). Interviewed in Paris, 9 July 2015 by EE.

IPCC WGII author and co-chair Chris Field (USA): Interviewed in Copenhagen, 20 August 2014 by EE, and in Paris, 8 July 2015 by EE and RK.

IPCC WGII author Saleemul Huq (Bangladesh). Interviewed in Paris, 9 July 2015 by EE.

IPCC WGII author Eystein Jansen (Norway). Interviewed in Oslo, 13 May 2015 by EE.

IPCC WGII author Sheila Jasanoff (USA). Interviewed in Paris, 9 July 2015 by RK.

IPCC WGII author and co-chair Hervé le Treut (France). Interviewed in Paris, 23 March 2015 by EE and again in Paris, 8 July 2015 by EE.

EE = Elisabeth Eide; RK = Risto Kunelius

aim of the interviews was to obtain a deeper understanding of how the IPCC works, the dilemmas faced by the scientists during the processes involved, and the challenges of communicating complicated scientific results to stakeholders.

THE IPCC: BACKGROUND AND CHALLENGES

The UN General Assembly endorsed the establishment of the IPCC in December 1988. The founding document addressed all UN-affiliated governments, asking them to immediately initiate "...action leading, as soon as possible, to a comprehensive review and recommendations" regarding climate change (UN 1988), and urging all UN sub-organizations to support this work. Although the primary goal was to provide a comprehensive scientific assessment of climate change, the IPCC was also established to translate *climate science* into *climate politics*: "...an attempt to create a new interface between science and policy, as a post-normal" *intergovernmental* and *scientific* organ (Hulme 2009: p. 92, see Chap. 1). Whereas "normal"

Sidebar 2.2

The IPCC

   The United Nations Environmental Programme (UNEP) and the World Meteorological Organization (WMO) established the Intergovernmental Panel on Climate Change[2] in 1988 to "[...] provide the world with a clear scientific view on the current state of knowledge on climate change and its potential environmental and socio-economic impacts".

   The IPCC's main products are the five Assessment Reports (AR) published thus far: AR1 (1988), AR2 (1995), AR3 (2001), AR4 (2007) and AR5 (2013–2014). The reports are divided into three working groups (WGs). WGI concentrates on the Physical Science Basis, WGII on Impacts, Adaptation and Vulnerability, and WGIII on the Mitigation of Climate Change. The IPCC also produces special Task Force reports on specific issues.

   The IPCC does not conduct original research; it assesses research by thousands of scientists from around the world (now 195 member countries). Scientists contribute their work on a voluntary basis. The *authors* (hundreds of experts selected by the WGs) do the actual science produces specific disciplinary knowledge, the IPCC aims to synthesize a global, hybrid form of knowledge from findings in many disciplines and areas of research. It also embodies an understanding of science as a *process* of constantly evolving assessment criteria and procedures rather than as an accumulation of *results*, determined by fixed practices (For a structural overview of the IPCC, see Sidebar 2.2).

The IPCC aims at being a global entity; however, it is worth noting that the developing world is structurally underrepresented. Valérie Masson-Delmotte, for instance, mentions that a number of countries cannot afford their own research centers, which weakens the acknowledgement of important issues, such as research on rain-fed agriculture in tropical Africa and Latin America. According to Saleemul Huq from Bangladesh, the global representation of scientists has improved since the AR3, but he still criticizes the current knowledge-power relations of research: "The science gets done in the North, and is then transferred to the South. This should be reversed, particularly in the case of adaptation where the 48 Least Developed Countries have done more and faster planning for adaptation than any country in the North. They [the LDCs] have also highlighted the South-South cooperation".[3]

The uneven global distribution of IPCC scientists may help explain the amount of media coverage generated by the IPCC reports. A shortage of local scientific sources can affect the ability of journalists to create a sense of *proximity* to the research disseminated. In Ugandan newspapers, for instance, the AR5 report launches appeared to be taking place in faraway places with hardly any nationals involved. In Brazil, on the other hand, journalists interviewed IPCC authors and highlighted papers written by local scientists that were mentioned in the reports. According to Chris Field, the IPCC has been "very ambitious about the regional, disciplinary, gender and seniority diversity" and hopes to build on this momentum in future reports. He adds that WGII included one researcher from Egypt and two from Bangladesh, but admits that representation from the Middle East, in particular, is weak.

The distribution of contributing authors is also uneven with respect to scientific discipline. In WGIII, according to IPCC author David G. Victor, nearly two-thirds of the 35 coordinating lead authors hailed from the field [of economics]; furthermore "...less than one-third of the 64 coordinating lead authors [in WGII] were social scientists, and about half of those were economists" (Victor 2015: p. 28). In other words, the participation of social sciences is weak and dominated by economists.

This volume addresses the issues of local variations and the communication challenges that the diffusion of IPCC messages face. In this chapter, however, we explore more in-depth the general science communication challenges faced by IPCC scientists, such as uncertainty and clarity, the non-prescriptive character of the IPCC, and some of the controversies surrounding this global institution.

IPCC Authors and Communication Challenges

Given the scale of climate change challenges, the relationship between science and common sense is particularly important. Climate change puts people's livelihoods and lifestyle choices on the global agenda. Thus "lay" understandings of science must be taken seriously:

> [...] it is risky to dismiss the non-scientific mentality a priori as ignorance in the face of uncertainty. Many scientific and technological developments are intrinsically uncertain, saturated by exuberance and imagination when the real impacts still need to be defined and monitored; and here public opinion is an asset, has a role to play and must be mobilized in controversial debates. (Bauer 2009: pp. 379–380)

Bauer's invitation to reconsider the boundary between science and common sense can also contribute to questioning the boundaries between different strands of science communication. Secko et al. (2013) have identified four useful science communication models. The *science literacy* model sees scientific facts and wisdom flowing to audiences and good science communication as a means of filling the audience's "knowledge deficit". The *contextual* model also focuses on the proper delivery of scientific information, but stresses the need to develop different criteria for different audiences. The *lay expertise* model acknowledges the limitations of science and highlights the role of science communication in the interaction between science and laypeople. Similarly, but even more radically, the *public participation* model sees science as deeply embedded in society and aims at democratizing science through more engaging communication (Secko et al. 2013: p. 6). These inherited and partly contradictory tendencies of science communication help to conceptualize some of the communication challenges recognized by IPCC scientists. In climate communication, models that rely on the authority of scientific knowledge (traditional and contextual) are faced with the question of how science can retain enough of the aura of certainty while communicating *probable* futures in a *risk-uncertainty* vocabulary. Some authority, after all, is sorely needed to argue for the need to act. The terrain of more dialogic models (lay expertise and participatory science) is not simple either. Sheila Jasanoff highlights this in her critique of the *stakeholder* notion, showing how being informed or not can influence a person's role:

> Even by calling people stakeholders, you are already pre-judging the situation. It is not obvious how somebody who does not yet see a stake [...] either because they are so disenfranchised that they do not see it, or because they are not good predictors, and they don't understand how the game will evolve so that it bears back on them. (Interview, 9 July 2015)

This stakeholder notion forms a key part of the IPCC rhetoric. It seems to refer mostly to politicians, who take ownership of the IPCC results. The "disenfranchised" are not part of this circuit of communication, although several of the IPCC authors interviewed mentioned the importance of civil society. Jasanoff adds "...we have a term, 'university-industry relations', but we don't have a comparable investment in university-civil society relations". Her critique resonates with a larger trend observed by Norman Fairclough (1993), in which the liberalization paradigm causes academic institutions to be more business-oriented in their external presentations. In such models, there is less space for ordinary citizens or particularly vulnerable groups (i.e. the disenfranchised).

The IPCC has faced its share of problems executing its demanding communication task. In 2009, a particularly influential public episode cast doubt on IPCC science, after a leak exposed thousands of emails and a number of documents produced by the Climate Research Unit (CRU) at East Anglia University. On the surface, the emails looked as if CRU researchers had been illegitimately tampering with evidence to strengthen the case for global warming. In the logic of the press, this event was quickly coined as "Climategate" (Pearce 2010). Despite the fact that in April and July of the following year, two reports cleared the CRU of serious allegations, the unit was criticized for its communication skills (Oxburgh 2010; Russell 2010). The so-called "Himalaya-gate", an erroneous prediction in IPCC AR4 that there was a "very high chance" that the Himalaya glaciers would melt by 2035, also tainted the IPCC's reputation. The misguided predictions were highlighted by the press in late 2009, and the IPCC leader, Rajendra Pachauri, had to back down and apologize for this claim in January 2010.[4] On the other hand, the "gate-ification" of climate science, in these cases, represents an unjust discursive twist through the association of climate researchers with the Watergate scandal and the illegal actions of US President Richard Nixon (1973).

Reflecting below on their experiences, IPCC scientists articulate the complexity of their work in mediating climate knowledge. Questions of *uncertainty* and *clarity* highlight problems with the traditional model of science communication. The challenges of being *policy relevant* (descriptive) but not *policy prescriptive* represent vital contextual (target audience) questions. We also address the *controversies* that have emerged during the IPCC processes.

Uncertainties as Obstacles?

Communicating uncertainty to a general audience has been a major obstacle for climatologists and authors of the IPCC reports (see Ekwurzel et al. 2011; Patt and Schrag 2003; Schenk and Lensink 2007; Jones 2011; Mastrandrea and Mach 2011). Experiments with public perception of the IPCC's probabilistic statements regarding uncertainty show that individual interpretations vary a lot and that a better language to describe uncertainty is needed (Harris and Corner 2011; Budescu et al. 2014; Cooke 2014).

At another level, the challenge for climate communication may also be *too much certainty*, or to be more exact, alleged certainty spurred by journalism's drive toward conclusiveness, drama and the sensational. Content analysis of a UK newspaper between 1987 and 2006 (Jennings and Hulme 2010), for instance, showed that journalists ignored the IPCC prediction that a certain kind of climate catastrophe is "unlikely" to happen. Instead, sensationalism dominated the coverage, representing the probability of a catastrophe as greater than its actual likelihood. Partly because of these earlier experiences, the IPCC revised its guidelines for AR5. Attention was drawn to the WGs through the use of specified uncertainty language, including consistent references to "…summary terms for evidence and agreement, levels of confidence, and quantified measures including likelihood" (Mastrandrea et al. 2011: p. 689). Valérie Masson-Delmotte is content that uncertainties were more accurately highlighted in AR5. This is important, she says, to constrain environmentalists from playing with fear and enhancing demobilization in the long run. Eystein Jansen adds that some of the uncertainties communicated at the AR5 report launches may have already changed, particularly concerning the reduced rate of warming over the past decade ("climate pause"): "at the time of publication, unfortunately the research working to explain this, was in progress, but not concluded. Much has happened after the publication". Here he alludes to recent findings that ocean heat content has been on a steady rise over the last decades and that excess heat has been transferred to the ocean interior, rather than remaining at the surface, particularly in the Pacific.[5]

Seita Emori comments on the uncertainty related to emissions and temperature rise: "We often speak about a *likely* probability if we refer to meeting the two-degree targets. It means 66 percent. [...] There is an uncertainty. In the course of five years or ten years of temperature and greenhouse gas emission, we might be able to narrow it down to 70". Thomas Stocker adds that scientists: "...have to be faithful and true to the assessment findings, which always include levels of confidence and assessments of the uncertainty. Even if we are using clear words, going back to the headline statements[6], it is always a Summary for Policymakers that is the basis of the conclusions and the communication. That consists of all the caveats, uncertainties, and numbers".

At the same time, there is an awareness that uncertainties may become less so as time passes. Jasanoff (2007) points out that uncertainty "... has become the threat to collective action", suggesting that the spread of binary thinking frames the future "...in terms of determinate choices between knowable options" (p. 34). Climate scientists also recognize problems linked to insufficient awareness of climate risks and the confusion it can create even among educated elites and policymakers: "Risk communication is now a major bottleneck preventing science from playing an appropriate role in climate policy" (Sterman 2011: p. 811).

CLARITY AND ITS LIMITS

Parallel to the problem of uncertainty runs the need to communicate with clarity. Albeit with certainty rising on anthropogenic causalities, climate scientists (will always) disagree on a number of issues. Their work is governed by sound skepticism toward simplified conclusions, not least since they treat future perspectives and options (Szerszynski and Urry 2010). As Hulme puts it, "Science thrives on disagreement. Science can only function through questioning and challenge. It needs the oxygen of skepticism and dispute in order to flourish" (2009: p. 75). This fundamental logic in the culture of science runs counter to the demands of traditional journalism and policymakers, who ask for clear-cut knowledge in order to perform their roles in society: to inform the public and to define the choices available.

According to Hervé le Treut the process through which science becomes journalism often leads to oversimplification and a focus on extremes: "The global results of the IPCC process are being used, and that is positive. Yet, in the process, [...] divergences occur between what we hear and what the scientific community is ready to say. For instance, there is over-attribution of extreme events to climate change, among politicians". Such a political and journalistic urge to connect the science to people's everyday lives may be well intended but it represents a turn away from the science itself. It reminds us of the *contextual* model of science communication, which retains a top-down approach that addresses scientific information in "specific audience-linked contexts" (Secko et al. 2013: p. 68). Acknowledging that science may take on different meanings in differing local contexts, this model "...constructs messages relevant to particular audiences while paying attention to the needs and situations of these audiences" (Secko et al. 2013: p. 68).

In journalism, the inclination to popularize scientific knowledge by finding illustrative "cases" may at best offer a way to build bridges between scientific and locally based knowledge, when it comes, for example, to farming or fishing. Such *clarity by exemplification* may have potential to reach larger audiences than (summaries of) scientific reports. Thus, as elaborated in Chap. 12, reflecting on good journalistic practices can sometimes solve the puzzle of clarity without jeopardizing scientific quality. On the other hand, overall clarity is in demand. "One thing is clear", says Thomas Stocker, "the two-degree target has requirements and these requirements were spelled out in WGIII very clearly". Whether journalists have the time and resources to digest the clarity and nuances of the AR5 remains a more problematic issue.

THE DESCRIPTIVE-PRESCRIPTIVE DILEMMA

The IPCC spells out requirements for climate action, but upholds its mandate by not indicating how they should be fulfilled, that is, by not being policy prescriptive. Governments acknowledge the *scientific* authority of the IPCC reports, but are not committed to any political implications these results may carry. According to the IPCC, "The work of the organization is therefore policy-relevant and yet policy-neutral, never policy-prescriptive".[7] Eystein Jansen admits that this policy-neutral attitude is not always easy, and that it can be more of a challenge in oral than in written communication. In lectures or interviews "...it is difficult not to have a more personal and subjective way of communicating. When you emphasize the dangers and the negative consequences, it should be okay to express a hope that negotiations will lead to some positive results". "It is easy to get rid of all the 'shoulds'", says Karen O'Brien, referring to the "likelihood" language. This "should" notion being associated with "pre-scriptive", which is not recommendable for IPCC scientists.

Thomas Stocker says that after 17 years of IPCC service the relevant/ prescriptive distinction is in his blood. However, there is a risk involved with journalists taking quotes out of context or shortening them:

> When I say, "Limiting climate change requires substantial and sustained reductions in greenhouse gas emissions" [...] If that well thought-through sentence, which is by no means policy prescriptive, but extremely policy rel-evant, is being abridged and shortened to say, "Stocker says that greenhouse gas emissions need to be cut by 70 percent", then, of course, that's beyond my control.

Saleemul Huq agrees that a scientist's main task is to refer to the Summaries for Policymakers (SPMs). He feels free, however, to give his personal opinion as long as it is clear that he speaks for himself: "We are as entitled as any citizen to give our own views". Several scientists make it clear that they view media as an invitation to play a more political role, and that even if unintended the result may be interpreted as policy prescriptive.

The way in which the IPCC is institutionally intertwined with world politicians is in itself a challenge to neutrality and to a non-prescriptive stance. Chris Field says that in Europe most politicians know a few scien-tists whom they trust and interact with at times. It is different in the USA where: "Most Americans don't know a scientist, and thus most of the time do not have such conversations. The climate change discussion [in the USA] is then driven by the political agenda. Very few of the conservatives ever speak of climate change". Field's observations point to the important role politicians play in fueling media attention. Jansen adds that the "... national agenda drives much of the climate change angles" and sees even more fundamental consequences: "...a substantial part of the [scientific] knowledge disappears, since *journalists are mostly interested in the messages that adhere to the national political agenda* or rivalries between political constellations" (emphasis added). On the other hand, the journalistic field could demonstrate its relative autonomy by focusing more on the science or by challenging political neglect.

SUMMARIES FOR POLICYMAKERS: SUITABLE COMMUNICATION?

The full reports from the IPCC AR5 (approx. 7000 pages) are not meant to be digested by the general public. The SPMs are short, 20- to 30-page distilled versions of the findings in each report. They represent a negotiated agreement between the authors in each WG and government representatives. The author groupings for an SPM consist of approximately 25 people; however, the group leadership has considerable power over the end result. Politicians rarely interfere with the scientific presentation of these results (see Controversies below). Karen O'Brien says the WG process "...is about making a lot of material consistent, while shortening. This does not necessarily make it more understandable for journalists". By contrast, the SPMs are *meant* to be accessible. According to Chris Field, "Leading scientists need to distill the results, so that decision makers themselves or aides are able to interpret them. The results should also be comprehensible for NGOs and [other] constituencies, in other words most leaders". Thomas Stocker is convinced that in the AR5 the IPCC overcame challenges represented by complexities to develop coherent messages.

> We have achieved statements like "human influence on the climate system is clear" or "limiting climate change requires substantial and sustained reductions in greenhouse gas emissions." We have succeeded to embed them fully as parts of the Summary for Policymakers, creating, thereby, ownership of these statements by the governments through the consensus.

He adds that none of the "headline statements" disappeared or were mutilated. But how far did these statements reach? In a critique of the SPMs, Richard Black (2015) states the obvious: "...[the] most important policymakers are not specialists in climate science or the economics of mitigation" (p. 282). Black believes the SPMs "...must be appropriately constructed for time-poor generalists who spend virtually all of their working lives outside the climate bubble". This statement is not only relevant to "stakeholders", but also to journalists, since the vast majority are generalists who cover climate change from that position. Black (2015) concludes that the SPMs "...are not necessarily attuned to the requirements of policymakers—or, indeed, to the general public who in the final analysis fund the IPCC" (p. 283).

Several IPCC authors underline that the SPMs are not meant for the public, but rather to provide politicians with a basis for their decision-making. Field, on the other hand, says that SPM authors tried to come up with a frame that could be used to communicate the report from the start. Moreover, they (at least the WGII report authors) tested their messages to "...figure out whether people got it or not. Our criteria were accuracy and understandability".

Eystein Jansen does not think that politicians even read the SPMs: "Rather, they receive a condensed version provided to them from their advisors, documents that are easier to use directly in their own communication". Several other interviewees agree, adding that the language is "extremely impenetrable for lay people", as Saleemul Huq iterates. Hervé le Treut thinks that "...even science students cannot read them. Very few seem to have read them. Then newspapers write a few lines". This is a rather precise description of the communication process from scientists to laypeople, who receive inadequate or limited information and thus remain more or less climate illiterate. Launch events and press releases—with condensed messages including quotes from interviewees—become starting points for many journalists, while few are likely to have read the full reports (Chap. 12).

Valérie Masson-Delmotte shares many of the same views and urges scientists to find other ways of communicating. Some IPCC representatives are "...heavily used for education and to raise the knowledge of the delegates to the UNFCCC [the COPs]". These extra media activities may represent an important addition to the (inadequate) communication of IPCC science. Jansen says it is "...easier to unambiguously formulate the results from natural sciences, since they are generally more concise regarding numbers and messages".

According to Saleemul Huq, the draft SPM receives "...comments from government bureaucrats [political level], a new draft is written and presented in plenary, where we go through line after line while the time pressure is great [...] If agreement is not reached, the leadership has some power and may relegate some formulations to footnotes". Huq notes that a bullet point on "loss and damages" in WGII was missing in the final version, "...even if this was taken up by lead authors from the very first year".

This competition for prominent space is an important characteristic of the WG processes. Jansen says nuances are left out of the SPMs, and that the language is "...calibrated through the negotiation process with the government representatives". According to him there is "...a struggle from the different chapters to have their conclusions reach the upper level". The meticulous monitoring of accuracy, language and meaning represents a particular challenge. Field illustrates this negotiation with a friendly quip targeting his colleagues: "From the English side we try to get rid of words, while the Germans add words—both to achieve clarity".

CONTROVERSIES

Crafting consensus around the SPM reports may be challenging. Controversies occur, such as the one on loss and damage mentioned above. Minutes from AR5 WGI discussions reveal that Saudi Arabia resisted the headline statement that the warming of the climate system is "...unequivocal, and since the 1950s, many of the observed changes are unprecedented over decades to millennia". Saudi Arabia claimed the statement was "alarmist" and urged that the terms "unequivocal" and "unprecedented" be qualified. It also requested use of the year 1850 instead of 1950 and called for a reference to slowed warming over the past 15 years. There were other divergent proposals, but in the end the original statement was approved "after some discussion".[8]

Another controversial incident came to light when, after the release of WGIII, one of the chapter authors publicly expressed his discontent. The author, Robert Stavins, was responsible for the final chapter of the SPM on "International Relations". In a blog post following the release, he placed blame not on co-authors or leaders but structural and institutional matters: "...as the week progressed, I was surprised by the degree to which governments felt free to recommend and sometimes insist on detailed changes to the SPM text on purely political, as opposed to scientific bases" (Stavins 2014). Frustrated by the process, he added: "In my view, with the current structure and norms, it will be exceptionally difficult, if not impossible, to produce a scientifically sound and complete version of text for the SPM on international cooperation that can survive the country approval process" (Stavins 2014). That such problems arose in a chapter on *international cooperation* is perhaps logical, since it has to do with UNFCCC negotiations (the COPs); national representatives approving of the SPM must be careful not to harm the negotiation positions of their respective governments.

The emphasis on economic growth became controversial when, during the approval process of the SPM for WGIII:

> [...] a small group of nations vetoed graphs that showed countries' emissions grouped according to economic growth. Although this format is good science—economic growth is the main driver of emissions—it is politically toxic because it could imply that some countries that are developing rapidly need to do more to control emissions. (Victor 2015: p. 28)

According to Jonathan Lynn, the IPCC Head of Communication and Media Relations, earlier public controversies about climate science emphasized the need for the professionalization of IPCC communication. In 2009–2010, there was no full-time communication person at the IPCC. Lynn, who has extensive journalistic experience, feels that by organizing media training sessions for IPCC scientists and (thus) creating a more systematic awareness of IPCC media coverage, they "...have turned around the media perception". Several of the IPCC researchers interviewed mention going through media training and characterize this as a positive experience.

CLIMATE PRAGMATISM: CONCERNING "THINGS PEOPLE LOVE"?

In *Can Science Fix Climate Change?* Mike Hulme (2014) challenges the scientific prospects involved in geoengineering a "global thermostat" and particularly the idea of "stratospheric aerosol injection"; what he calls "techno-fixing the climate". He rejects this for three main reasons: controlling global temperature does nothing toward controlling local weather; the model is ungovernable and unattainable; and there may be a series of unintended consequences that "...would lead to unending experimentation" (Hulme 2014: pp. 118–119). He does not reject all efforts to curb the rising global temperature (the two-degree target, highlighted by the IPCC), but states that climate change needs to be framed and addressed in a different manner. In short, he endorses a view requiring "greater humility with regard to the limits of scientific knowledge than is implied by planetary engineers—and also a greater exercise of compassion and justice in the deployment of that knowledge" (Hulme 2014: p. xiv). What he calls for is "climate pragmatism". This entails addressing people's needs by creating adaptive measures to make local communities more resilient to the ongoing threats of climate change. He suggests curbs on all climate gases that are not CO2 (which will have more immediate effects), and to meet future energy demands through massive political commitment and investment in eco-friendly technology. His book also challenges mainstream framings of climate change and its "solutions" as something which can be fixed at the top level, either in political negotiations through the COPs or by transnational conglomerates of technical expertise.

How do Hulme's suggestions apply to the ways in which journalists communicate the IPCC science—and to the challenges of making them relevant to and understood by ordinary people? Communicating climate change as *risk* has been one of the guidelines for AR5 (see also Chap. 5) to meet these challenges: "Risk communication is now a major bottleneck preventing science from playing an appropriate role in climate policy" (Sterman 2011: p. 811). Another problem may be "self-censorship". Antilla's study of the media coverage of "tipping points" mentions examples of delayed information of nuclear radiation experiments in the USA due to "Cold War values" as well as examples where fear of "alarmism" could lead to underreporting an issue associated with political or environmental agendas.

Beck, on the other hand, says global climate change "has been 'brought home', especially in the West, as possibly 'the' global risk of the age" (2010: p. 261). He highlights the ways news media have adopted a performative stance by using visually arresting images so "...the abstract science of climate change is rendered culturally meaningful and politically consequential; geographically remote spaces become literally perceptible, 'knowable' places of possible concern and action" (Beck 2010: p. 261). This is a fairly optimistic interpretation of the effect of dramatic images, as they may also have the reverse effect, inspiring visions of doom and gloom (Hahn et al. 2012; Eide 2013).

Hervé le Treut opens up the same discussion by referring to his own experiences in the Southwestern region (Aquitaine) of France, where he was surprised to witness "...huge participation [in climate change discussions] by a variety of scientists and experts" as well as local producers (fishermen and wine farmers). Furthermore, local media are involved. He also "...found that people were interested *since it was about things they love, the sea, the mountains, etc.* They were worried that climate change would change the places they love" (emphasis added). According to Le Treut, the local dimension "also reconciles adaptation and mitigation". He adds that the regional aspect would reconcile researchers "...with communities who think we work purely in abstract ways—they never liked the way we globalize things".

Valérie Masson-Delmotte admits to being "people-oriented". As a scientist (glaciologist) she must speak for the climate science community, but she sometimes volunteers for school programs on outreach events about climate, and has written books for children about climate change. On the other hand, the IPCC "cannot frame their SPMs as if they were aimed at non-specialists", says IPCC communication director Jonathan Lynn and adds that scientists still react to "media framing", but also that "a lot of language and presentations still go over the top of the heads". Lynn tells of how he once challenged a scientist to present his findings in ten minutes. "He was horrified, but did it." This episode in a simple way demonstrates the dilemmas in science communication. Scientists spend years conducting field research and data analysis, but the science is then "brought down to size" in an SPM, and furthermore by media. When asked how he treats invitations to speak within the frame of a TV news item lasting approximately 90 seconds, Jansen says he mostly accepts such offers: "Usually such a short item is linked to the situation and the news. You have to try and find some pegs. Of course, in such a situation one appears with more certainty than in the research material".

On the other hand, accuracy is a burning issue in science, not least when it comes to quantifications concerning the future. In journalism, nuances may disappear. Hervé le Treut says that when speaking on the science one cannot violate scientific ethics and rules, but must stay "true to these ideals. Many journalists do not understand this. We talk about measuring 4.2 or 4.6 [as quantifying temperature rise], but they think it is so much easier to understand the figure 6, so they put that". In such a case, exaggeration may add to the drama, but not to human engagement.

How IPCC scientists relate to media raises the question of how to frame the information in new ways. Field, while iterating that the IPCC can go on stating that they were right ten years ago and five years ago, must now frame their findings with more emphasis on solutions. He feels that the solutions space "presents some opportunities for new kinds of coverage".

THE POLITICS OF FRAMING

With less than adequate training in climate science, the simple journalistic way to tackle climate change is to relegate the topic to the political field, concentrating on negotiations regarding what efforts should be taken to mitigate global warming. In the process of translating matters of fact into matters of concern, the temptations of "doom and gloom", or the *alarmist* perspective, are rampant. Swyngedouw (2010) labels this tendency an "ecology of fear", writing that "'Fear' is indeed the crucial node through which much of the current environmental narrative is woven" (p. 217). Furthermore, he claims that

> [...] sustaining and nurturing apocalyptic imaginaries is an integral and vital part of the new cultural politics of capitalism [...] for which the management of fear is a central leitmotif. [...] At the symbolic level, apocalyptic imaginaries are extraordinarily powerful in disavowing or displacing social conflict and antagonism. (Swyngedouw 2010: p. 218)

As a basis for his argument, the author mentions the "rapidly growing derivatives market" and the "commodification of CO_2" (2010: p. 220). A critic of the post-political perspective, Swyngedouw (2010) then elaborates on "...the predominance of a managerial logic in all aspects of life, the reduction of the political to administration where decision-making is increasingly considered to be a question of expert knowledge and not of political position" (p. 225). Berglez and Olausson (2014) refer to Swyngedouw in their explorative study on how "...the post-political condition of climate change is established in public discourse" (p. 15). Their findings indicate that perceptions of climate change in everyday experiences and practices "... generate a consensual discourse on anthropogenic climate change which, however, does not challenge the *singular Cause* of climate change—the capitalist system" (Berglez and Olausson 2014: p. 15). As demonstrated in other chapters, attempts at profound systemic critique *does occur* in the IPCC coverage, although not as a dominant discourse. Thus, it matters enormously how climate is *framed*. Discussing this, Lakoff (2010) rejects the *conservative* moral system as unfit to deal with global warming due to its idea of man being above nature, its market-decides-all ideology and its disregard for systemic causation. He contrasts this with a *progressive* moral system (Lakoff 2010) that includes empathy, responsibility and the ethic of excellence. In addition, Ytterstad's (2014) research shows that when asked anonymously scientists, to a greater extent than journalists, reply that fundamental systemic changes must occur if we are to mitigate global warming.

Precisely how engaged journalists and editors are in reporting climate science is another question. In mainstream journalism, one of the structural challenges is how to adequately train for the job. There are few climate specialists working within the journalist corps in any given country. Chris Field mentions a handful in his own country (the USA) and adds that "... they do a good job, but need to fight to reach the front page and not page nine". On the other hand, Saleemul Huq has registered a genuine and growing media interest in climate change in Bangladesh. He is highly sought after as a columnist and frequently contributes to the Bangladeshi press. His advice to editors is to invest reporter time on this topic, and he urges journalists to "...find your nearest climate scientist and build a relationship".

Karen O'Brien's advice to reporters is to "try to present a different perspective. The same narrative is much on replay. Journalists must help readers to connect the dots, for instance when it comes to climate change and the drought in California". She also thinks that there may be a psychological need among people to downplay the bad odds and thus a "positive spin" might deserve a space. On social media, O'Brien says she finds many positive stories that are generating lots of response, suggesting: "There is an appetite for positive news".

Conclusion: Navigating between Constraints

To journalism, whose responsibility it is to communicate science to the general public, Secko et al. (2013) lay expertise, and public participation models seem particularly relevant as ways to empower local populations by regarding their experiences as important. These people-oriented models resonate with Jasanoff's "policies based on humility". This is exemplified by way of valuing "...greenhouse gases differently depending on the nature of the activities that give rise to them; and uncover[ing] the sources of vulnerability in fishing communities before installing expensive Tsunami detection systems" (2007: p. 33). Scientists within the IPCC framework, on the other hand, mostly practice the "science literacy model" due to structural and other reasons, while in their capacity as citizens, they may feel free to choose more participatory modes of communication. In this context, the social sciences, still waiting to be more prioritized by the IPCC structures, may contribute to a more open debate. Victor suggests important reforms, including the IPCC becoming a "...more attractive place for social-science and humanities scholars" and structural changes in the shape of a "supporting process" more independent from governmental influence that helps to address questions "...the IPCC cannot handle on their own". Among these questions are relations between policies and emissions, and "…how different concepts of justice and ethics could guide new international agreements that balance the burdens of mitigation and adaptation" (Victor 2015: p. 29).

Simon Cottle appears to share some of these concerns when he writes that the nature of global crises today "…invites a deeper appreciation of how ecological crises and environmental conflicts, though *ontologically* rooted in the complex forces of globalizing late modernity, are nonetheless *epistemologically* dependent on how they become discursively framed, visualized and dramatized in media and communication" (2013: pp. 28–29). He also calls for "global journalism" and "increased research attention in the years ahead" (Cottle 2013: p. 32). This research should take into account both the interconnectivity between "old" and "new" media, and attempt to bring "ontological and epistemological outlooks closer together" (2013: p. 29). The AR5, in its wider scientific approach (bringing in more disciplines), tried to do this at times, but, as the interviews in this chapter indicate, there are doubts concerning how well the essential scientific conclusions and their perspectives were digested by stakeholders and society at large. The scientists represented underline that their main outreach (the SPMs) is directed at policymakers, not the general public. Several interviewees doubt whether politicians and laypeople are even capable of digesting the SPMs, even if they were made more accessible than the previous Assessment Reports.

Another emerging doubt is whether journalistic institutions are inclined to prioritize this global crisis, in an era where these institutions themselves are facing major setbacks and increased commercialization pressure. We may also ask to what extent we can expect a "global journalism" at a moment when news media are cutting back on foreign correspondents and on specialized editorial staff. Nevertheless, as emphasized later in this volume (Chap. 12), there are reporters eager to maintain these perspectives, and as this chapter demonstrates, there are certainly scientists who feel an urgency to reach beyond institutionalized stakeholders in their communication and who appreciate the role of other actors (NGOs and popular movements) in mitigating climate change.

The challenges are formidable. A study of laypeople's views on climate change revealed that the involved participants mostly considered climate change (and its societal importance) "an intractable problem, alluding to uncertainties in the science and the ineffectiveness of actions as major constrains to addressing the issue" (Lorenzoni and Hulme 2009: p. 391). This ineffectiveness may well be attributed to the political stakeholders to whom the IPCC reports are addressed, and to their lack of will to be "policy prescriptive" in a way that several IPCC scientists suggest they would like to see but do not openly acknowledge.

In a recent paper addressing the AR5 communication experiences, IPCC communication director, Jonathan Lynn (2016a), juxtaposes science and media interests.

Scientists are trained to describe their findings by marshalling large amounts of evidence and building to a conclusion. This is the opposite to communication in media, which in its simplest form starts with a conclusion, supports it with context and one or a few outstanding pieces of evidence and backs it with interpretation that is more emotional or colourful.[9]

He adds that this is how people are familiar with receiving information via the media. As demonstrated above, this does not reduce the challenges for climate scientists, communicators or journalists in particular. IPCC scientists are constrained by their loyalty to the scientific results, while also being partners in deliberations with state actors from around the world. In addition, there is the urge to enhance climate literacy among citizens. When it comes to navigating in this landscape of concerns, the responses of scientists are nuanced. Some strictly adhere to the non-policy prescriptive position, while others do so less, acting as individuals. Some communicate extensively with media and citizens and thus contribute to the popularization of IPCC science, while others take positions that are less public. However, generally speaking, we can now see that the IPCC has put communication on the agenda to a much larger extent than in the past, as a recent event exclusively dedicated to IPCC communication issues clearly attests.[10] This speaks to the urgency of the matter, and to the fact that media coverage of climate science is often left to journalists without the expertise to interpret the IPCC reports, even in a condensed format.

One way to improve the situation is to engage more in training the journalists who cover climate change. When asked in a project led by James Painter, scientists ranked training for general journalists and "more journalists specializing in climate science" as the two most important ways to improve climate reporting (2010: p. 82). Another solution is perhaps for scientists to engage in what Secko et al. (2013) call the *lay expertise* model, allowing for more interaction between science and citizens. Here journalists, whose (ideal) role is to represent citizen interests, can play the role of intermediaries.

Notes

1. IPCC. (2007). *Climate Change 2007: Synthesis report*. R. K. Pachauri & A. Reisinger (Eds.). Geneva, Switzerland: IPCC. Retrieved from https://www.ipcc.ch/publications_and_data/ar4/syr/en/mains2-4.html
2. IPCC. (2015). *Organization*. Geneva, Switzerland. Retrieved from http://www.ipcc.ch/organization/organization.shtml
3. Huq, S. (2015, July). *Keynote address*. Our common future under climate change, International Science Conference, Paris, France.
4. Pearce, F. (2010). Claims Himalayan Glaciers could melt by 2035 were false, says UN scientist. *The Guardian*, January 20. Retrieved from http://www.theguardian.com/environment/2010/jan/20/himalayan-glaciers-melt-claims-false-ipcc
5. See, for example, http://www.epa.gov/climatechange/science/indicators/oceans/sea-surface-temp.html
6. IPCC. (2013). Headline statements from the summary for policymakers. *Climate Change 2013: The physical science basis*. Geneva, Switzerland: IPCC. Retrieved from http://www.climatechange2013.org/images/uploads/WG1AR5_Headlines.pdf
7. IPCC. (2015). *Organization*. Geneva, Switzerland. Retrieved from http://www.ipcc.ch/organization/organization.shtml
8. Earth Negotiations Bulletin. (2013, September 23). *Summary of the 12th session of Working Group I of the Intergovernmental Panel on Climate Change*. Retrieved from http://www.iisd.ca/vol12/enb12581e.html
9. Lynn, J. (2016a). IPCC communications issues—Constraints and opportunities. *Advance paper submitted to the IPCC Expert Meeting on Communication*. Oslo, Norway: IPCC. Retrieved from https://ipcc.ch/meeting_documentation/pdf/Communication/160119_advance_paper_on_constraints-JLynn.pdf
10. For a summary of the meeting and archive of the papers presented, see: IPCC. (2016, February 9–10). *Meeting report of the Intergovernmental Panel on Climate Change Expert Meeting on Communication*. J. Lynn, M. Araya, Ø. Christophersen, I. El Gizouli, S. J. Hassol, E. M. Konstantinidis, K. J. Mach, L. A. Meyer, K. Tanabe, M. Tignor, R. Tshikalanke, & J. P. van Ypersele (Eds.). Geneva, Switzerland: IPCC. Retrieved from http://www.ipcc.ch/meeting_documentation/meeting_documentation_ipcc_workshops_and_expert_meetings.shtml

Bibliography

Austgulen, M. H., & Stø, E. (2013). Norsk skepsis og usikkerhet om klimaendringer. *Tidsskrift for samfunnsforskning, 54*(2), 124–150.
Bauer, M. W. (2009). Editorial. *Public Understanding of Science, 18*, 378–382.

Beck, U. (2010). Climate for change or how to create a green modernity? *Theory Culture & Society, 27*(2–3), 254–266.

Beckett, C. (2010). *The value of networked journalism.* London: London School of Economics and Political Science.

Berglez, P., & Olausson, U. (2014). The post-political condition of climate change: An ideology approach. *Capitalism Nature Socialism, 25*(1), 54–71.

Black, R. (2015). No more summaries for wonks. *Nature Climate Change, 5,* 282–284.

Boykoff, M. T., Daly, M., Gifford, L., Luedecke, G., McAllister, L., Nacu-Schmidt, A., et al. (2015). *World newspaper coverage of climate change or global warming, 2004–2015.* Center for Science and Technology Policy Research, Cooperative Institute for Research in Environmental Sciences, University of Colorado. Retrieved from http://sciencepolicy.colorado.edu/icecaps/research/media_coverage/world/index.html

Budescu, D. V., Por, H., Broomell, S. B., & Smithson, M. (2014). The interpretation of IPCC probabilistic statements around the world. *Nature and Climate Change, 4,* 508–512.

Capstick, S., Whitmarsh, L., Poortinga, W., Pidgeon, N., & Upham, P. (2015). International trends in public perceptions of climate change over the past quarter century. *WIREs Climate Change, 6,* 35–61.

Cooke, R. M. (2014). Messaging climate change uncertainty. *Nature Climate Change, 5,* 8–10.

Cottle, S. (2013). Environmental conflict in a global media age: Beyond dualisms. In L. Lester & B. Hutchins (Eds.), *Environmental conflict and the media* (pp. 19–37). New York: Peter Lang.

Eide, E. (2012). Visualizing a global crisis. Constructing climate, future and present. *Conflict and Communication Online, 11*(2), 1–16.

Ekwurzel, B., Frumhoff, P., & McCarthy, J. (2011). Climate uncertainties and their discontents: Increasing the impact of assessments on public understanding of climate risks and choices. *Climatic Change, 108*(4), 791–802.

Fairclough, N. (1993). Critical discourse analysis and the marketization of public discourse: The universities. *Discourse and Society, 4*(2), 133–168.

Hahn, O., Eide, E., & Ali, Z. S. (2012). The evidence of things unseen: Visualizing global warming. In E. Eide & R. Kunelius (Eds.), *Media meets climate: The global challenge for journalism* (pp. 217–242). Gothenburg: Nordicom.

Harris, A. J. L., & Corner, A. (2011). Communicating environmental risks: Clarifying the severity effect in interpretations of verbal probability expressions. *Journal of Experimental Psychology Learning, Memory & Cognition, 37*(6), 1571–1578.

Hulme, M. (2009). *Why We Disagree About Climate Change: Understanding Controversy, Inaction and Opportunity.* Cambridge: Cambridge University Press.

Hulme, M. (2014). *Can science fix climate change? A case against climate engineering.* Cambridge: Polity Press.

IPCC. (2013). Stocker, T. F., Qin, D., Plattner, G.-K., Tignor, M., Allen, S. K., Boschung, J., Nauels, A., Xia, Y., Bex, V., & Midgley, P. M. (Eds.), *Climate change 2013: The physical science basis. Summary for policymakers.* Contribution of Working Group I to the Fifth Assessment Report of the Intergovernmental Panel on Climate Change. Cambridge: Cambridge University Press.

IPCC (2015). Structure: How does the IPCC work? Geneva, Switzerland. Retrieved from http://www.ipcc.ch/organization/organization_structure.shtml

Jasanoff, S. (2007). Technologies of humility: Citizen participation in governing science. *Nature, 450*(7166), 33.

Jennings, N., & Hulme, M. (2010). UK newspaper (mis-)representations of the potential for a collapse of the thermohaline circulation. *Area, 42*(4), 444–456.

Jones, R. (2011). The latest iteration of IPCC uncertainty guidance-an author perspective. *Climatic Change, 108*(4), 733–743.

Lakoff, G. (2010). Why it matters how we frame the environment. *Environmental Communication, 4*(1), 70–81.

Lee, T. M., Markowitz, E. M., Howe, P. D., Ko, C.-Y., & Leiserowitz, A. A. (2015). Predictors of public climate change awareness and risk perception around the world. *Nature Climate Change, 5,* 1014–1020.

Lorenzoni, I., & Hulme, M. (2009). Believing is seeing: Laypeople's views of future socio-economic and climate change in England and Italy. *Public Understanding of Science, 18,* 383–400.

Lynn, J. (2016a). IPCC communications issues—Constraints and opportunities. *Advance paper submitted to the IPCC Expert Meeting on Communication.* IPCC, Oslo, Norway. Retrieved from https://ipcc.ch/meeting_documentation/pdf/Communication/160119_advance_paper_on_constraints-JLynn.pdf

Mastrandrea, M., & Mach, K. (2011). Treatment of uncertainties in IPCC Assessment Reports: Past approaches and considerations for the Fifth Assessment Report. *Climatic Change, 108*(4), 659–673.

Mastrandrea, M. D., Mach, K. J., Plattner, G., Edenhofer, O., Stocker, T. F., Field, C. B., Ebi, K. L., & Matschoss, P. R. (2011). The IPCC AR5 guidance note on consistent treatment of uncertainties: a common approach across the working groups. *Climate Change 108,* 675–691.

Nerone, J. (2015). *The media and public life: A history.* Cambridge: Polity.

Oxburgh Report. (2010, April). *Summary.* Norwich, UK. Retrieved from http://www.uea.ac.uk/documents/3154295/7847337/SAP.pdf

Painter, J. (2010). *Summoned by science: Reporting climate change at Copenhagen and beyond.* Oxford, UK: Reuters Institute for the Study of Journalism.

Patt, A. G., & Schrag, D. P. (2003). Using specific language to describe risk and probability. *Climatic Change, 61*(1/2), 17–30.

Pearce, F. (2010). *The climate files: The battle for the truth about global warming.* London: Guardian Books.

Russell Report. (2010, July 7). *The independent climate change e-mails review*. London, UK. Retrieved from http://www.cce-review.org/

Schenk, N. J., & Lensink, S. M. (2007). Communicating uncertainty in the IPCC's greenhouse gas emissions scenarios. *Climatic Change, 82*(3/4), 293–308.

Secko, D., Amend, E., & Friday, T. (2013). Four models of science journalism: A synthesis and practical assessment. *Journalism Practice, 7*(1), 62–80.

Stavins, R. N. (2014, April 25). Is the IPCC government approval process broken? An economic view of the environment. A blog. Retrieved from http://www.robertstavinsblog.org/2014/04/25/is-the-ipcc-government-approval-process-broken-2/

Sterman, J. (2011). Communicating climate change risks in a skeptical world. *Climatic Change, 108*(4), 811–826.

Swyngedouw, E. (2010). Apocalypse forever? Post-political populism and the spectre of climate change. *Theory, Culture & Society, 27*(2–3), 213–232.

Szerszynski, B., & Urry, J. (2010). Changing climates: Introduction. *Theory, Culture and Society, 27*(2–3), 1–8.

UN General Assembly Resolution 43/53. (1988, December 6). Protection of global climate for present and future generations of mankind, A/RES/43/53. United Nations General Assembly: 70th Plenary Meeting. Retrieved from http://www.un.org/documents/ga/res/43/a43r053.htm

Victor, D. G. (2015). Embed the social sciences in climate policy. *Nature, 520,* 27–29.

Ytterstad, A. (2014). Vite, men ikke røre? Profesjonelle grenser for klimahandlinig hos norske forskere og journalister. In E. Eide, D. Elgesem, S. Gloppen, & L. Rakner (Eds.), *Klima, medier og politikk*. Oslo: Abstrakt forlag.

CHAPTER 3

Attention, Access and the Global Space of Interpretation: Media Dynamics of the IPCC AR5 Launch Year

Risto Kunelius and Dmitry Yagodin

The publication of the IPCC AR5 reports in 2013 and 2014 offers a unique case study for capturing the role of journalism in global communication. As the IPCC's years of work were condensed first into the full reports (WGs) and then the Summaries for Policymakers (SPMs), enormous amounts of information and scientific insight were put into action at the moment of publication. This moment initiated a partially routinized and partially unexpected process that mobilized a transnational infrastructure of communication. Identifying some of the key features of this structure and understanding its dynamics is the core target of this study. In this chapter, we provide a first view of what happened at the global level as the findings of scientists and the shared interpretations of the IPCC's scientific-political actors were compressed through the needle hole of publicity.

R. Kunelius (✉)
School of Social Sciences and Humanities, University of Tampere, Tampere, Finland

D. Yagodin
School of Communication, Media and Theatre, University of Tampere, Tampere, Finland

© The Author(s) 2017
R. Kunelius et al. (eds.), *Media and Global Climate Knowledge*,
DOI 10.1057/978-1-137-52321-1_3

59

Social theory offers ample definitions about what constitutes a "public" as well as the peculiar characteristics of such social formations (e.g. from Habermas 1989 to Warner 2002 to Fraser 2014). While this is not the place to rehearse the nuances of this debate, a shorthand version of some key conceptual issues is useful to shape the questions of this chapter.

Publics and publicity are made of *attention* and *access.* Primarily they are conceived by the attention of an audience. It is attention—the ability to momentarily address a large number of people—that opens up a space of discourse for social actors to appear in. Audience attention, and the potential it offers to inform and persuade, makes this public space an important site of power struggles. Access to this space—and control over what and who is represented in it—is the key practice of exercising power and influence through the public. In real life moments of publicity, power over attention (what is talked and thought about) and access (who has the privilege or duty to perform in public) are deeply intertwined, but analytically, they offer two different measures of the public. First, we can think of public attention as a measure of the *importance* of focusing on an issue or problem, and access as a measure of the amount of *power* possessed by the actors at hand. However, the elusiveness of the topic returns when we are presented with findings and conclusions on who and what gets discussed in the public space opened up by journalism. There is, after all, always the possibility that *not* having attention focused on particular issues and *not* being invited to speak publicly serve particular interests too. Thus, it is perhaps more accurate to say that the ability to *control* attention and access is what is at stake when it comes to relations of power in public space.

This chapter uses this shorthand theory of publicity to analyze the launch of the IPCC AR5 report. We look at four episodes—the publication of the three Working Group (WG) reports and the Synthesis Report (SYR)—during a yearlong event that began in December 2013 and ended in December 2014. We shed light on the global dynamics of this long act of publication by tracking the shifts in newspaper attention during this year and the distribution of actors in the spaces of interpretation opened by this attention. To better understand the broad dynamics between the IPCC and the media, we also look at the way the panel tried to build key messages into its press releases and analyze the main themes that broke through to the media at the level of story headlines. It is worth pointing out that—quite strikingly, given the media saturated era of politics we live in—the AR5 cycle was the first time in the IPCC's history that professionally crafted press releases, produced by the panel *itself*, were used to communicate the results (Lynn 2016).

ATTENTION, HIGHLIGHTS AND VOICES

The analysis in this chapter unfolds by combining four kinds of empirical evidence. Figure 3.1 provides an overview of the *general amount of media attention* paid to each WG report. The attention score displayed in the graph combines the amount and size of all stories which were predominantly focused on climate and made clear reference to the IPCC ($N = 780$).[1] The figure highlights two important things. First, there were significant differences in media attention between the 22 countries studied. We offer Brazil (high) and China (low) as examples here to provide a sense of this range. Such differences, of course, are a core part of the broader study, and we will tackle them from different perspectives throughout the book (in particular, see Chaps. 4–5, 7–8 and 10). Second, and more importantly for the argumentation here, we identify how the amount of media attention fluctuated from one WG report to another. *WGI: The Physical Science Base* made the most prominent headlines in December

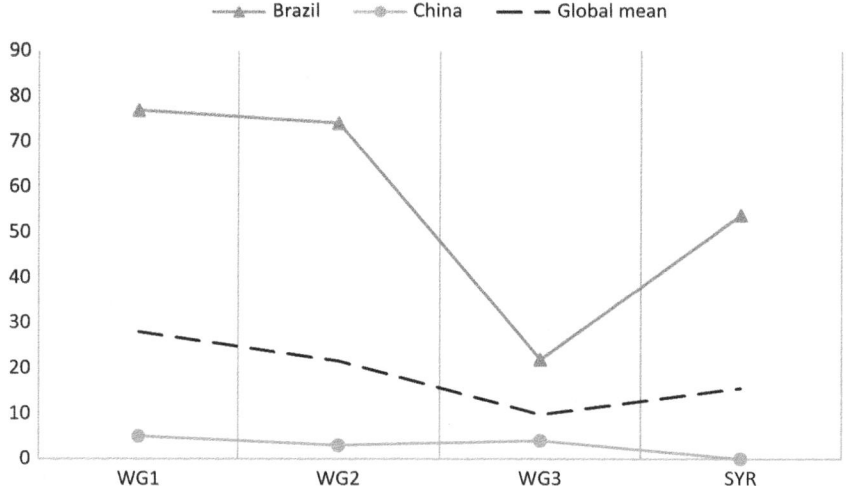

Fig. 3.1 Media attention to AR5 working group reports

---

[1] For more details on this sub-sample, see Chap. 1. For a more elaborated discussion of the headline coding methodology, see the Appendix.

Table 3.1 Headline highlights of media attention to different working group reports (percentage in $N = 551$)

	WGI	WGII	WGIII	SYR	Total
Science/evidence	33.2	8.6	4.5	5.4	16.7
Global concern	33.2	43.4	43.3	40.2	39
General action	13.6	22.9	43.3	35.9	24.2
Morality	12.1	18.9	23.9	19.6	17.1
Politics	25.6	13.7	13.4	15.2	18.4
Local concern	13.1	28.6	0	7.6	15.6
Specific action	5	13.1	38.8	14.1	13.5

Note: One headline can refer to several highlight aspects.

2013; *WGII: Impact, Adaptation and Vulnerability* did not quite reach the same level in March 2014. In April 2014, global media attention dropped rather low for the *WGIII: Mitigation of Climate Change*, before gaining some strength in December 2014 with publication of the SYR.

Attention, by definition, must point to something: content or a theme inside the broader field of climate change and politics. In order to capture this, Table 3.1 offers a rough analysis of the *thematic content* of the headlines in published stories around the world. Here we use a slightly smaller sample of stories, focusing on stories where the AR5 reports were the main topic ($N = 551$).[2] We then juxtapose these findings with a reading of the concrete output (press releases) of the IPCC. The results of these *headline highlights* provide only a rough, proximate sense of the dominant contents that captured the media's attention. The thematic analysis is based on coding the headlines from a secondary material where headlines, originally published in several languages, were translated into English by coders on the research team.[3]

The following seven themes were identified as the most salient ones in the headlines, and the thematic content was subject to further analysis:

---

[2] See Chap. 1 for details on the sample.
[3] The coding scheme was developed through an initial, iterative reading of all the headlines. Then, three coders individually coded the translated headlines with discrepancies discussed and solved through negotiation. It is also worth noting that one headline could be coded to several themes.

- *Scientific evidence*—headlines that emphasized scientific evidence but did not link these to consequences or the need to act. Discussion on the uncertainty of the evidence and blunt factual statements or conclusions about man-made climate change were coded here.
- *Global concern*—headlines that emphasized climate change as a problem that the whole world must face. Such references often had an alarmist tone.
- *General action*—headlines that operated on the same scale, but also underscored the demand for action (mitigation or adaption).
- *Morality*—headlines that mobilized some kind of reference to justice, responsibility or other moral arguments, either in relation to time (future generations), location (South–North) or power hierarchies (politicians–people). Explicit use of "we" or references to "blame" were coded here.
- *Politics*—headlines that emphasized differences and controversies, such as those referring to skepticism (politicizing the IPCC), or references to tensions, disputes or disagreements between stakeholders.
- *Local concern*—headlines that translated these general concerns into concrete problems and challenges, either geographically or topically.
- *Specific action*—headlines that identified more concrete measures for action in specific geographic areas or policy domains.

The overall results show a tendency for journalism to emphasize and frame climate change and the IPCC as a story of generalized, global concern. They also suggest a relative decline in the focus on issues of scientific evidence and political controversy during the AR5 launch year.

Journalism is always co-produced by journalists and their sources. To capture this, Table 3.2 shows the distribution of the *voices of news actors* during each WG launch ($N = 780$). Here, again, we use the same sample as the attention graphs. The analysis of voices operationalizes the question of access.[4] It illustrates the dominance of scientific actors in the coverage, highlighting the way journalists gave room to scientific authority. It also shows clearly the overall importance of domestic science actors in making sense of the reports in the media. Finally, the rise of political voices (both transnational and national) is noteworthy, reflecting a shift in the dominance of the interpretive space from science to politics as the launching year progressed toward the SYR.

Below, we look at each episode of the WG report publications more closely, considering the ways in which the IPCC message was framed and

[4] See Chap. 1 for details on the voice coding.

Table 3.2 Voices overview (percentage of total AR5 sample 780 stories, 1332 voices)

	WGI	WGII	WGIII	SYR	All
National politics	17.4	11.4	17.6	23.6	17.2
Transnational politics	11.2	7.2	5.1	20.5	11.4
Civil society	14	18.6	17.6	19.2	16.9
Business	2.2	3.6	3.4	4.4	3.2
Domestic science	32.9	30	27.3	20.5	28.6
Foreign science	18.2	25.6	26.7	8.8	19.2
Media	4	2.5	2.3	2.7	3.1
Other		1.1		0.3	0.4

constructed. This analysis contributes to earlier research on IPCC media coverage (O'Neill et al. 2015; Painter 2013, 2014; Hulme 2009a) by providing an overview of a particularly wide, global sample.

While our wide geopolitical reach adds to past research, it is important to acknowledge that the object of analysis in this chapter is an artificially constructed one. The "global attention" dynamics presented here are based on analysis of a sample of stories that never caught the attention of a specific, real audience. The "public" that this sampling and analysis constructs, then, is a theoretical one. As a measure of average global reaction, it provides a rough indicator against which to discuss the IPCC's communication strategies and the reactions of global coverage during 2013–2014.

The Physical Science Base (WGI): Evidence Confirmed and Questioned

The first AR5 sub-report focused on the basic ontological and epistemological issue of climate change: questions concerning whether and how fast human-caused climate change is happening and what we know about the past, present and future of climate change. In coverage of WGI, then, the question of *scientific evidence* was strongly highlighted in headlines (Table 3.1). There were also distinct signs of an emphasis on "political" aspects of the issue (bringing forth potential for explicit disagreements). This suggests that in 2013 journalism expected another round of debate on the epistemological politics of climate change ("Is climate change really happening?" "Is it really man-made?"), between the IPCC and the different varieties of climate skeptics. In several countries, discussions about the "climate pause" made headlines before the WGI launch. *Aamulehti* in Finland communicated conflict, "Controversial climate report: warming has slowed" (26 September 2013); while Swedish *Dagens Nyheter* headlined: "Not trustworthy about likelihood" (27 September 2013), echoed by *Svenska Dagbladet*: "Climate panel surrounded by controversy" (27 September 2013). More moderately, a *Toronto Star* (Canada) headline wore that "Scientists debate rate of warming caused by CO2; 'Climate sensitivity' a key factor in setting emission limits" (27 September 2013). There were also leaks highlighting the dramatic conclusions of WGI, such as Norway's *Dagsavisen* whose headline read "The renewal age has to start now" (27 September 2013).

Sidebar 3.1
WGI press release excerpts

It is extremely likely that human influence has been the dominant cause of observed warming since the mid-twentieth century. The evidence for this has grown, thanks to more and better observations, an *improved understanding* of the climate system response and *improved climate models*.

Warming in the climate system is *unequivocal* and since 1950 many changes have been observed *throughout the climate system* that are *unprecedented* over decades to millennia. Each of the last three decades has been successively warmer at the Earth's surface than any preceding decade since 1850.

"[...] Our assessment of the science finds that the atmosphere and ocean have warmed, the amount of snow and ice has diminished, the global mean sea level has risen and the concentrations of greenhouse gases have increased," said Qin Dahe, Co-Chair of IPCC Working Group I.

Thomas Stocker, the other Co-Chair of Working Group I said: "Continued emissions of greenhouse gases will cause further warming and changes in all components of the climate system. Limiting climate change will require *substantial and sustained reductions of greenhouse gas emissions*."

(*continued*)

Sidebar 3.1 (continued)

"Global surface temperature change for the end of the twenty-first century is projected to be likely to exceed 1.5°C relative to 1850 to 1900 in all but the lowest scenario considered, and likely to exceed 2°C for the two high scenarios," said Co-Chair Thomas Stocker. "*Heat waves* are very likely to occur more frequently and last longer. As the Earth warms, we expect to see currently *wet regions receiving more rainfall*, and *dry regions receiving less*, although there will be exceptions," he added.

"[...] global mean sea level will continue to rise, but at a faster rate than we have experienced over the past 40 years," said Co-Chair Qin Dahe.

[...] Co-Chair Thomas Stocker concluded: "As a result of our past, present and expected future emissions of CO2, we are *committed to climate change, and effects will persist for many centuries* even if emissions of CO2 stop."

Source: IPCC. (2013, September 27). Human influence on climate clear, IPCC report says. *IPCC Press Release*. Retrieved from http://www.ipcc.ch/news_and_events/docs/ar5/press_release_ar5_wgi_en.pdf (emphasis added).

Leaks and pre-launch news about a possible "climate pause" exemplify media savvy attempts by networks of skeptics to *preframe* the public debate on the AR5 report before it had a chance to define its own public footing. Here, the unavoidably slow official process of accepting the final version of the report kept the IPCC silent until the public AR5 launches. This gave a momentary advantage to voices outside the panel who spoke ahead of the publication. Thus, skeptics and other stakeholders were given a chance to revive the image of a less unified science community. In spite of their weak scientific grounds, claims about the "climate pause" provided some journalistically useful angles (controversy) for the media. While explicitly ignoring the pre-publication debate, the WGI press release (27 September 2013) emphasized the *increasing certainty* of scientific conclusions on the human influence on warming (Sidebar 3.1). The release underlined the weight of evidence and predicted the consequences of climate change in a broad, global framework.

The press release, thus, provided an implicit answer to the "climate pause" argument by referring to ocean temperatures. However, it did not address these leaks explicitly since this is not the IPCC's role (Painter 2014: pp. 47–52). It also underlined both the extreme weather events and the fact that climate change is indeed taking place and will occur no matter what we do ("We are committed to climate change").

The message of WGI came across rather well in the immediate media reaction sampled. Over half of the voices in the stories (Table 3.2) came from scientists. At the headline level, journalists read the report as *confirmation* that climate science was now even more certain than before. The sense of alarm was globally prevalent, not least in Bangladesh: "Alarm rings for climate system" (*The Daily Star*, 28 September 2013) and "IPCC's new report: Increasing warming: Human beings are more liable" (*Prothom Alo*, 28 September 2013). This combination of evidence and alarm was also highlighted in *Sydney Morning Herald* (Australia): "Evidence is there, so what are we waiting for?" (28 September 2013), and with softer wording in the British *Telegraph*: "Global warming unequivocal, say scientists" (28 September 2013). *People's Daily* (China) drew attention to melting glaciers and rising sea levels (28 September 2013), while Japan's *Yomiuri Shimbun* attributed extreme weather to global warming (28 September 2013). Some newspapers focused more on the global political consequences. *The Ethiopian Herald's* headline read "IPCC urges world to reduce greenhouse emission" (29 September 2013), while *Le Monde* in France (perhaps already anticipating the Paris COP process and the central role of France) concluded: "Even more pressure on international negotiations" (28 September 2013).

While the thematic *news* reception of WGI largely followed the IPCC warnings and consensus, *opinionated* coverage, particularly the op-ed sections in some newspapers, dedicated some space to suspicions and doubts concerning the trustworthiness of IPCC predictions. In our sample, *The Telegraph* (UK) and *Business Day* (South Africa) both seemed to keep the door open for skeptical voices on their opinion pages. Two of these headlines explicitly accuse the IPCC of being politically biased, while others question the science.

IMPACT, ADAPTATION AND VULNERABILITY (WGII): SOCIAL AND OTHER CONSEQUENCES

Global media attention declined somewhat for coverage of WGII (Fig. 3.1). Headline highlights (Table 3.1) show a small rise in journalistic attempts to make climate change more concrete and tangible: locally identified concerns and more specifically defined actions were present at the headline level of media attention. In the big picture, this offered a balancing counterweight to the dominant themes of global and general concern. At the same time, headlines were less likely to emphasize the "political" (disagreements and negotiations).

Changes in the general level and tone of attention can be viewed as a modest success for the WGII communication strategy. The headline on its press release reads "IPCC Report: A changing climate creates pervasive *risks* but *opportunities* exist for effective responses. Responses will face challenges with high warming of the climate" (31 March 2014, emphasis added). The press release (Sidebar 3.2) develops a dialectic between high concern (risks) and the potential for future-oriented action (opportunities). It emphasizes that "adaptation" is not only a passive and reactive agenda but must become more future-oriented, and in doing so open new horizons for action.

Sidebar 3.2
WGII press release excerpts
    The report concludes that responding to climate change involves *making choices about risks* in a changing world. [...] The report identifies *vulnerable people, industries, and ecosystems around the world*. It finds that risk from a changing climate comes from vulnerability (lack of preparedness) and exposure (people or assets in harm's way) overlapping with hazards (triggering climate events or trends).
    [...] "Climate-change adaptation is not an exotic agenda that has never been tried. Governments, firms, and communities around the world are building experience with adaptation," Chris Field [WGII Co-Chair].
    [...] "With high levels of warming that result from continued growth in greenhouse gas emissions, *risks will be challenging to manage*, and even serious, sustained investments in adaptation will face limits," said Field.
    [...] "The report concludes that people, societies, and ecosystems are vulnerable around the world, but with *different vulnerability in different places*. Climate change often interacts with other stresses to increase risk," Field said.

(*continued*)

Sidebar 3.2 (continued)

[…] Field added: "Understanding that climate change is a challenge in managing risk opens a wide *range of opportunities* for integrating adaptation with *economic and social development* and with initiatives to limit future warming. We definitely face challenges, but understanding those challenges and tackling them creatively can make climate-change adaptation *an important way to help build a more vibrant world* in the near-term and beyond."

Source: IPCC. (2014, March 31). IPCC Report: A changing climate creates pervasive risks but opportunities exist for effective responses. *IPCC Press Release*. Retrieved from http://www.ipcc.ch/pdf/ar5/pr_wg2/140330_pr_wgII_spm_en.pdf (emphasis added).

As previous studies have suggested, the use of an elaborated and effective language of "risk" could be seen as a potentially effective way of engaging audiences (Painter 2013, 2015b). The rhetoric of *warnings* combined with a belief that *risks can be managed* helps link climate change to tangible, more concrete issues and policy choices (Table 3.1). Many news headlines focused on the consequences for human beings and other species. In the UK, for example, *The Guardian* headlined: "The threats: Vision of the future: poor and marginalized are least to blame but will suffer the most" (31 March 2014). *O Estado de São Paulo* in Brazil wrote: "Decrease in bee populations will affect agriculture" (29 March 2014). Finnish *Helsingin Sanomat* (30 March 2014) predicted "More storm damages, more expensive insurance," while Egyptian *Daily News* stated "Climate change boosts conflict, risk, floods, hunger" (31 March 2014). *The Toronto Star* declared that Asia would be the hardest hit with climate change to "displace millions" (31 March 2014). Notably, the more positive side of the WGII framing—climate adaptation as a "way to help build a more vibrant world"—rarely made the headlines.

The change in media reception between WGI and WGII reflects differences in their mandates and expertise. WGII, which comprises a combination of climate science and social sciences, is bound to focus on (negative) local and more specific consequences of climate change. This concretization of the topic—in the shape of displacement, poverty rise and ecological changes—was clearly evident in the headlines. Looking at the distribution of voices, however, we can also point to the low level of national political voices in WGII coverage (the lowest of all WGs, see Table 3.2). This suggests that the space of public interpretation opened by risk management and adaptation is not yet a lucrative one for political actors.

MITIGATION OF CLIMATE CHANGE (WGIII): ALARM AND HOPE

Global media attention sank considerably for WGIII in most of the 22 countries sampled (Fig. 3.1). The relative rise of "specific action" in the headline analysis (Table 3.1) reflects the ways in which headlines took up specified emission cut targets (mostly on the global scale, rarely at the national level). While less focus was placed on *how* these targets could be reached (means of mitigation), journalists clearly used the report to make the abstract global concern about climate change more tangible and real. Many headlines highlighted WGIII's demanding mitigation targets (reduce emissions 40 to 70 percent by 2050). They made this effort in

---

Sidebar 3.3

WGIII press release excerpts

A new report by the Intergovernmental Panel on Climate Change (IPCC) shows that global emissions of greenhouse gases have risen to *unprecedented levels despite a growing number of policies to reduce climate change*. Emissions grew more quickly between 2000 and 2010 than in each of the three previous decades.

According to the Working Group III contribution to the IPCC's Fifth Assessment Report, *it would be possible*, using a wide array of technological measures and changes in behaviour, to limit the increase in global mean temperature to two degrees Celsius above pre-industrial levels. *However, only major institutional and technological change will give a better than even chance* that global warming will not exceed this threshold.

[…][WGIII Co-Chair Ottmar] Edenhofer said. "There is a clear message from science: To avoid dangerous interference with the climate system, *we need to move away from business as usual*."

Scenarios show that to have a likely chance of limiting the increase in global mean temperature to two degrees Celsius, means lowering global greenhouse gas emissions by 40 to 70 percent compared with 2010 by mid-century, and to *near-zero by the end of this century*.

(continued)

Sidebar 3.3 (continued)

Ambitious mitigation may even require removing carbon dioxide from the atmosphere.

[…] *Estimates of the economic costs of mitigation vary widely.* In business-as-usual scenarios, consumption grows by 1.6 to 3 percent per year. Ambitious mitigation would reduce this growth by around 0.06 percentage points a year. However, the underlying estimates do not take into account economic benefits of reduced climate change.

[…] Slowing deforestation and planting forests have stopped or even reversed the increase in emissions from land use. Through afforestation, land could be used to draw carbon dioxide from the atmosphere. This could also be achieved by combining electricity production from biomass and carbon dioxide capture and storage. However, as of today this combination is not available at scale, *permanent underground carbon dioxide storage faces challenges* and the risks of *increased competition for land need to be managed.*

[…] "Climate change is a global commons problem," said Edenhofer. "International cooperation is key for achieving mitigation goals. *Putting in place the international institutions needed for cooperation is a challenge in itself.*"

Source: IPCC. (2014, April 13). IPCC: Greenhouse gas emissions accelerate despite reduction efforts. Many pathways to substantial emissions reductions are available. *IPCC Press Release*. Retrieved from http://www.ipcc.ch/pdf/ar5/pr_wg3/20140413_pr_pc_wg3_en.pdf (emphasis added).

spite of the fact the suggested emission reduction numbers were daunting—and difficult to understand in terms of their practical consequences. At the same time, in the headlines, the mitigation report drew increasing attention to moral responsibility. The overview of news actor voices for WGIII (Table 3.2), in turn, suggests a relatively weak position for transnational actors, possibly reflecting the strong national political interests related to mitigation issues. Noteworthy, throughout the WGs but in WGIII especially, where issues of economic growth and the interests of different branches of industry are forcefully implicated (both as problems and solutions to mitigation), is the small number of business community voices that journalists brought into the interpretative space (Table 3.2). From a global perspective, this is perhaps one of the weakest aspects of journalism's performance. Unlike NGOs and civil society actors, business actors may not need to be a visible part of this coverage.[5] This shows that journalists did not manage to engage them very effectively in the public space of interpretation.

WGIII presented a more demanding communication task than the two earlier reports. The core message of its launch (Sidebar 3.3) underscored the AR5's main theme: that time for effective action is indeed running out. While WGII could (at least rhetorically) present some hopeful opportunities, the evidence of mitigation trends and the scenarios developed for AR5 presented WGIII authors with fewer occasions to invoke the dialectic of alarm and hope. The press release made an effort: "IPCC: Greenhouse gas emissions accelerate despite reduction efforts. Many pathways to substantial emissions reductions are available" (13 April 2014). However, beneath this weak optimism lies reality and evidence: effective global mechanisms of reducing emissions have not been found. Indeed, as the press release for WGIII shows, actual emissions are growing at a pace that corresponds to "business as usual" predictions (leading to severe temperature rise), while political debates are centered on the global target average of two degrees. Thus, the report illustrated quite sharply the discrepancy between the public discourse of climate governance and politico-economic choices. The challenge—*start* by cutting 40–70 percent of emissions in the coming decades—is daunting in itself. The fact that this demands technologies that are still not widely available adds an additional counterweight to optimism. This also lays bare the *depth* of the climate challenge (Chap. 1): a key factor behind the emission rise is the current mode of economic growth, the very factor that is supposed to pay for mitigation investments. "Business as usual" should not be an option. Despite the immediate lack of attention in our global sample, the way the WGIII conclusions were communicated played a crucial role in shaping climate policy discourse building toward the 2015 Paris Agreement (for an insider assessment of the WGIII communication effort, see Sokona et al. 2016).

Newspaper headlines found different ways of dealing with this "Optimism with a touch" as Norway's *Aftenposten* (perhaps ironically) coined it (15 April 2014). China's *People's Daily* launched this general appeal: "Implement national strategy for adapting to climate change" (15 April 2014). Some took up—often as a positive option—local concerns.

---

[5] On efforts to target this community directly, see Wittington (2016).

Brazil's *O Estado de São Paulo* wanted to enhance the role of the country's own products: "Bio fuels will be 'critical', says IPCC" (13 April 2014). In Russia, *Kommersant* (online) did the same: "UN experts invite investments into gas industry for the sake of climate" (18 April 2014). *The Telegraph* (UK) considered the benefits of fracking: "Fracking could help to avoid climate change" (14 April 2014). Some headlines endorsed more radical options and lessons: the need to leave fossil fuels in the ground (*The Telegraph*, 12 April 2014), or to enhance green energy sources ("Triple clean energy output, say UN experts," *The Guardian*, 12 April 2014).

Newspapers often solved the discrepancy of hope and gloom by making weak allusions to hope in the headlines. This was exemplified by a *Sydney Morning Herald* headline which reads "Cost of safer, greener future could be less than is feared" (14 April 2014) or in *The Daily Star* (Bangladesh): "World can meet UN climate goal: UN panel urges 40–70 percent cut to greenhouse gas emissions by 2050" (14 April 2014). *Kompas* in Indonesia chose a techno-optimist headline: "Technology solves climate change" (15 April 2015). Some, such as the Finnish *Aamulehti*, chose a more dire perspective: "Despite measures, more and more emissions" (14 April 2014). Germany's *Süddeutsche Zeitung* opted for more systemic change: "IPCC: We will have to change our way of doing business" (14 April 2014); as did South Africa's *Cape Times*: "Soaring greenhouse gas emissions pile of pressure for change in practices" (14 April 2014).

Several factors may explain the decline of overall media attention from WGII to WGIII (see also O'Neill et al. 2015: p. 382; Painter 2014: pp. 38–39). It may be a result of the short-term attention logic of journalism. Scheduling the WG launches almost back to back in the spring of 2014 meant that WGII ate away at the attention potential and newsworthiness of WGIII. This plausible factor suggests that while there was some difference in the way media received WGI and WGII, both launches were strongly interpreted thematically within the framework of global concern (Table 3.1). Thus, while WGII clearly created new emphasis on local and concrete social concerns, it also managed to reiterate as "news" in March 2014 what had already been stated in September 2013: the increasing urgency of the IPCC's overall diagnosis. Stating the same fundamental message again for WGIII did not work. O'Neill et al. (2015: p. 382) also suspects that the WGIII contents were partly too technical for media and lacked compelling narratives (available in science controversies [WGI] and human suffering [WGII] for instance). Related to narratives and ideology, then, one may raise a more fundamental explanation. Perhaps the lack of attention to mitigation challenges is a troublesome sign of public denial:

Sidebar 3.4

Synthesis Report press release excerpts

Human influence on the climate system is **clear and growing,** with impacts observed on all continents. If left unchecked, climate change will **increase the likelihood of severe, pervasive and irreversible impacts** for people and ecosystems. **However, options are available** to adapt to climate change and implementing stringent mitigations activities can ensure that the impacts of climate change remain within a manageable range, creating a brighter and more sustainable future.

[...] said R. K. Pachauri, Chair of the IPCC. "The solutions are many and allow for continued economic and human development. **All we need is the will to change**, which we trust will be motivated by knowledge and an understanding of the science of climate change."

[...] The Synthesis Report makes a clear case that many risks constitute **particular challenges for the least developed countries** and vulnerable communities, given their limited ability to cope. People who are socially, economically, culturally, politically, institutionally, or otherwise marginalized are especially vulnerable to climate change. Indeed, **limiting the effects of climate change raise issues of equity, justice, and fairness** and is necessary to achieve sustainable development and poverty eradication. [...]

[said Vicente Barros, WGII Co-Chair]. "Adaptation is so important because it can be integrated with the pursuit of development, and can help prepare for the risks to which we are already committed by past emissions and existing infrastructure."

But adaptation alone is not enough. Substantial and sustained reductions of greenhouse gas emissions are at the core of limiting the risks of climate change. [...]

[...] "It is technically feasible to transition to a low-carbon economy," said Youba Sokona, Co-Chair of IPCC Working Group III. "But what is lacking are appropriate policies and institutions. The longer we wait to take action, the more it will cost to adapt and mitigate climate change."

Source: IPCC. (2014, November 2). Concluding installment of the Fifth Assessment Report:

Climate change threatens irreversible and dangerous impacts, but options exist to limit its effects. *IPCC Press Release*. Retrieved from http://www.ipcc.ch/pdf/ar5/prpc_syr/11022014_syr_copenhagen.pdf (emphasis added).

the media is not very good at suggesting fundamental changes to the way of life it is—itself—structurally based upon. The *no more business-as-usual* frame WGIII demanded was difficult to turn into headline coverage.

THE SYR: GLOBAL CONCERN

The SYR saw the overall attention level rise from the dip of WGIII, although it did not quite reach the level of the WGI launch. In many developing countries, the SYR was the most covered report (Chap. 4), but in the global overview WGI and SYR were the main targets of media attention (Fig. 3.1). The distinct difference between them, however, was that in the SYR coverage, national and transnational political actors increasingly took over the mediated interpretative space. Civil society voices were also prominent, in particular transnational NGOs (Table 3.2). This shows how the political stakes increased as the AR5 launch reached its conclusion and global climate politics turned toward the Paris COP in December 2015.

The press release for the SYR (2 November 2014), as one would expect, drew together the results of the three WGs. It also retained the earlier (in the WGs) established dialectical rhetoric between dire warnings and opportunities for action: "Concluding instalment of the Fifth Assessment Report: Climate change threatens irreversible and dangerous impacts, but options exist to limit its effects" (Sidebar 3.4).

This overall drift from focusing on scientific evidence (and controversy) to global concern and moral (justice) overtones was also reflected in the headlines (Table 3.1). *Kompas* in Indonesia wrote: "The inaction of developed countries is highlighted" (12 November 2014), and *Svenska Dagbladet* in Sweden mirrored the troubling future from the perspective of the developing world: "Tuvalu cannot handle many more degrees" (4 November 2014). While the politicians dominated as actors in the news space, this did not seem to "politicize" the main frames (headlines) of the stories. Rather, newspapers conveyed a strong sense of common, global urgency. *The Daily Star* in Bangladesh declared "Time running out for 2C target" (3 November 2014); *Business Day* in South Africa took a global position too: "UN: Time is against us for climate action" (3 November 2014). In Poland, the *Rzeczpospolita* made the hopeful claim "The world can do without oil and coal" (4 November 2014). In Germany, *Die Tageszeitung* eventually asked, "Is the clean-up going to start?" (15 November 2014), and in France, *L'Humanité* drew dramatic conclusions "How to respond to the climate emergency? A renovated governance" (13 November 2014).

REVIEWING THE BROAD GLOBAL RESPONSE

In this chapter, we have taken a first measure of the global space of interpretation that was opened up by mainstream journalism in 22 countries as a response to the four AR5 report launches in 2013–2014. In order to sketch the concrete interplay between IPCC communication output and the journalistic reaction, we also looked at the way the main press releases for each of the WGs framed their key message. Such condensation of the findings has been challenging for the panel. It has been a matter of principle until AR5 that the panel merely lays out the collected and synthesized evidence. Thus, its communication strategy has been undeveloped, relying on intermediaries, such as NGOs and individual scientists, to come up with sharper interpretations for public use. Practically, the task is a challenging one. The SPMs, which condense the much lengthier official reports, undergo a formal approval process (by scientist and governments) at a detailed linguistic level. This, in turn, makes it difficult for IPCC actors to simplify the language into a more accessible style and reading level. However, during AR5 press releases were crafted jointly by the co-chairs of each WG (content) in collaboration with the IPCC communication team (Lynn 2016), with new kinds of media training and preparation organized to support this new endeavor. Despite this effort, in early assessments of the IPCC AR5 communication strategy, several critical evaluations tackle the question of editing the language of science for more general public use, both in terms of comprehensibility and effectiveness (Barkemeyer et al. 2015; Painter 2015b; Hassol 2015).

A first look at the global attention dynamics shows that WGI, on the physical science basis, was initially received with some echoes of the old "controversies" of climate science. However, the IPCC's overall message in the first launch worked fairly well. This is no doubt a result of the enhanced communication efforts. In its first sub-report, for instance, WGI prepared a two-page "headline statements" analysis. This condensed version of the SPM was accepted through the same procedure as the SPM, and thus reflected the officially sanctioned position of the panel.[6] Condensing hundreds of pages first into the SPM and then even further into a two-page document—while sustaining a consensus among scientists and governments—is quite a remarkable achievement in itself.

---

[6] IPCC. (2013). *Climate Change 2013. The physical science basis* (WGI). Headline Statements from the Summary for Policymakers. Retrieved from http://www.ipcc.ch/report/ar5/wg1/

WGII, on vulnerability and adaptation, assumed a slightly different strategy. The report achieved relatively high media attention, and its core framing—the dialectic of risk and opportunity—resulted in the rise of more concrete and tangible questions (although the language of risk did not break through entirely, see also Chap. 5).[7] At the headline level, abstract global concern remained high and the headline coverage profile resembled WGI.

The global attention level sank quite dramatically for the WGIII report on mitigation. While the drop can be explained by a journalistic "climate fatigue," it also raises more fundamental questions than merely the critique of bad timing with the back-to-back launches of WGII and WGIII. Grim mitigation targets and an extremely challenging message concerning the speed of necessary global emission cuts can be seen as reason for a partial *denial*—that is, a subtle downplaying of the immediate news value of the report. Whereas WGII was able to contrast risks with new opportunities and change narratives, the story line of WGIII was left trying to convince its readers that radical emission cuts were possible. The SYR saw a new rise in attention levels and brought in more transnational, political and NGO voices. This reflects the somewhat successful role of the IPCC in working toward a new global climate agreement at the 2015 Paris COP.[8]

Journalism interprets the world by distributing attention and highlighting particular aspects of a topic. By doing so, a space opens that invites actors to appear and make sense of the object of attention. Explaining and elaborating on an issue, thus, is always a co-productive moment between journalists and their sources; "parajournalists" as Schudson (2003) aptly characterizes the routine sources of news. The interpretative space of AR5 coverage was—quite expectedly—dominated by scientists themselves. In the AR5 sample ($N = 780$, see Chap. 1), scientists dominated the news space

---

[7] WGII did not produce a "headline statement" document. Instead it included a similar, condensed summary of "top-level findings." This document addressed several adaption dimensions in concrete terms, from fresh water resources to health issues and economic costs of adaptation. See IPCC. (2014). *Impact, adaptation and vulnerability.* Top-level findings from the Working Group II AR5 summary for policymakers. Retrieved from: http://www. ipcc.ch/report/ar5/wg2/

[8] In addition to enhancing public communication efforts, IPCC AR5 knowledge was integrated into preparations for the Paris COP. See Fischlin, A. (2016, February 9–10). A new science-policy interface. The structured expert dialogue of the 2013–2015. Review of the UNFCCC. *Advance paper for IPCC Expert Meeting on Communication,* Oslo, Norway. Retrieved from https://www.ipcc.ch/meeting_documentation/pdf/Communication/ Structured_Expert_Dialogue_2013-2015_AFischlin.pdf

throughout the launch year (Table 3.2). Inside science, *natural scientists* were overwhelmingly dominant, with social scientists and economists playing only a marginal role. This was also the case with WGII and WGIII, even if the themes of these reports clearly pointed to a wider range of scientific expertise. This reflects the disengagement of other scientists in the IPCC launch efforts and, as a consequence, the prevalent journalistic interpretation of climate change as a predominantly natural science problem (see also Chaps. 2 and 13). *Domestic scientists* were clearly quoted more often than foreign ones, with about half of them explicit members of the IPCC author collective. As we shall see later, the role of domestic scientists is a crucial factor in the performance of journalists globally (Chaps. 4 and 10). Journalists and scientists, especially local scientists from various disciplines, could do much more to increase the weight and diversity of coverage.

Journalism's global performance in mobilizing other voices and stakeholders shows a strong role for national political actors. This suggests, again, not surprisingly, that the language of national interests operates as the main interpretative frame for thinking about necessary and possible climate action. The voices of domestic national politicians were more common than foreign ones, although toward the SYR report, this began to balance out. This hints at journalism trying to take a more global perspective on the issue and increase pressure for transnational political conclusions to be drawn. Moreover, not just foreign national voices but also *transnational* political voices (albeit always marginal compared to national actors) gained strength toward the SYR-launch. The increased emphasis on political actors toward the end of the AR5 process, however, was not met with increased *politicization* in the headline analysis, where the politics of *controversy* and *disagreement* did not increase notably. Albeit a weak signal, this suggests that global journalism too favored a "post-political" (Chap. 1) discourse on climate change. Formulated as a commonly shared sense of political urgency, directed toward politicians as a powerful elite, the narrative strayed away from actively opening up differences that could politicize the matter more concretely.

In the global reaction, civil society voices were mostly dominated by NGOs, with foreign NGOs gaining strength toward the end of the AR5 process. Yet the role of ordinary people was also fairly high, pointing at least in a rudimentary sense, to a weak tendency for journalists to incorporate some lay expertise and a sense of the broader, public-participation-oriented models of science communication. Interestingly, this access was clearly restricted to domestic ordinary people. Recognition of *foreign laypeople* as voices in the space of interpretation was marginal, echoing Jasanoff's critique on the importance of defining who is and who is not a "stakeholder" (Chap. 2). Silencing the distant laypeople can be seen as an example of the power of structural exclusion ("misframing" some key "affected" people from the debate, to follow Fraser's (2014) wording, see Chap. 6). In a critical vein, the meager role demanded from and awarded to business actors—even in reports on adaptation and mitigation, with all their risks and opportunities—is hard to justify journalistically. In total, business actors accounted for only 1 percent of the voices in the AR5 sample. This may be because these actors were not actively seeking to be part of this coverage. However, it also shows that journalists did not manage to engage them very effectively.

In judging the performance of global journalism here, it is good to bear in mind that the examined materials represent a fairly well resourced, elite-oriented media sample. This slice of mainstream journalism represents newsrooms that would, within the institutional boundaries of professional journalism, be best prepared to provide good coverage of the AR5 and its conclusions. Thus, given the amount of time (a year, at least) that journalists at these leading news organizations had to prepare for the AR5, and given that journalists must have *known* that not much new information was coming, their ability to effectively draw on a diversity of voices, and a diversity of scientific disciplines, leaves much to be desired.

BIBLIOGRAPHY

Barkemeyer, R., Dessai, S., Monge-Sanz, B., Gabriella Renzi, B., & Napolitano, G. (2015). Linguistic analysis of IPCC summaries for policymakers and associated coverage. *Nature Climate Change, 6,* 311–316.

Fraser, N. (2014). Transnationalizing the public sphere: On the legitimacy and efficiency of public opinion in a post-Westphalian world. In K. Nash (Ed.), *Transnationalizing the public sphere* (pp. 8–42). Cambridge: Polity.

Hulme, M. (2009a). Mediated messages about climate change: Reporting the IPCC fourth assessment in the UK print media. In T. Boyce & J. Lewis (Eds.), *Climate change and the media* (pp. 117–128). London: Peter Lang.

IPCC. (2013). Stocker, T. F., Qin, D., Plattner, G.-K., Tignor, M., Allen, S. K., Boschung, J., Nauels, A., Xia, Y., Bex, V., & Midgley, P. M. (Eds.), *Climate change 2013: The physical science basis. Summary for policymakers.* Contribution of Working Group I to the Fifth Assessment Report of the Intergovernmental Panel on Climate Change. Cambridge: Cambridge University Press.

IPCC. (2014, March 31). *IPCC report: A changing climate creates pervasive risks but opportunities exist for effective responses.* Geneva, Switzerland: IPCC. Retrieved from http://www.ipcc.ch/pdf/ar5/pr_wg2/140330_pr_wgII_spm_en.pdf

O'Neill, S. J., Kurz, T., Williams, H. T., Wiersma, B., & Boykoff, M. (2015). Dominant frames in legacy and social media coverage of the IPCC Fifth Assessment Report. *Nature Climate Change, 5,* 380–385.

Painter, J. (2013). *Climate change in the media: Reporting risk and uncertainty.* Oxford, UK: I.B. Tauris and Reuters Institute for the Study of Journalism.

Painter, J. (2014). *Disaster averted? Television coverage of the 2013/14 IPCC's climate change reports.* Oxford, UK: Reuters Institute for the Study of Journalism.

Painter, J. (2015b). The effectiveness of the IPCC communication: A survey of (mainly) UK-based users. *Advance paper submitted to the IPCC Expert Meeting on Communication.* IPCC, Oslo, Norway. Retrieved from http://www.ipcc.ch/meeting_documentation/meeting_documentation_ipcc_workshops_and_expert_meetings.shtml

Schudson, M. (2003). *The sociology of news.* New York: Norton & Co.

Sokona, Y., Edenhofer, O., Pichs-Madruga, R., Eickemeier, P., & Minx, J. (2016). Communicating the science of climate change mitigation: AR5 experiences from Working Group III. *Advance paper submitted to the IPCC Expert Meeting on Communication.* IPCC, Oslo, Norway. Retrieved from http://www.ipcc.ch/meeting_documentation/meeting_documentation_ipcc_workshops_and_expert_meetings.shtml

Warner, M. (2002). *Publics and counterpublics.* New York: Zone Books.

Whittington, E. (2016). Translating IPCC AR5 to business audiences. *Advance paper submitted to the IPCC Expert Meeting on Communication.* IPCC, Oslo, Norway. Retrieved from http://www.ipcc.ch/meeting_documentation/meeting_documentation_ipcc_workshops_and_expert_meetings.shtml

CHAPTER 4

Mediated Civic Epistemologies? Journalism, Domestication and the IPCC AR5

Risto Kunelius and Dmitry Yagodin

> Climate change ironically results from the three great success stories of the twentieth century: capitalist pursuit of profit, nation-states' commitment to economic growth, and citizens' pursuit of mass consumption rights. To challenge them is to challenge the three most powerful institutions of recent years. (Mann 2013: pp. 11–12)

Tension between the local and the global is a constitutive factor of journalism. On the one hand, journalistic culture is naturally inclined toward the global: it carries strong historic links to commerce and transportation, a fascination with technology and the mobility of people, curiosity about all that is new, and an affinity to abstract modern values, such as "freedom" or "progress." On the other hand, journalism remains deeply anchored to the immediate communities it aims or pretends to serve. Local languages, tastes, moral norms and common sense constitute the ingredients with which journalism

R. Kunelius (✉)
School of Social Sciences and Humanities, University of Tampere, Tampere, Finland

D. Yagodin
School of Communication, Media and Theatre, University of Tampere, Tampere, Finland

© The Author(s) 2017
R. Kunelius et al. (eds.), *Media and Global Climate Knowledge*,
DOI 10.1057/978-1-137-52321-1_4

imagines it's "public." Journalism is thirsty for the concrete and the detailed, the here and now, and it is dependent on local news sources and institutions of power. In particular, the history of modern journalism runs parallel to the formation of the nation-state (Anderson 2006; Thompson 1995; Sonwalkar 2008; Nerone 2015). The idea of the nation, the networks of power and the bureaucracies of the nation-state have provided a naturalizing framework for journalism to represent itself as an institution that pursues the common good. The construction of the national "we" has enabled journalism to negotiate a working compromise with its internal local–global tension. As a source of solidarity, the nation-state has appeared modern enough for journalists to challenge their local communities (regions, cities, municipalities) in the name of progress. At the same time, the idea of nation has provided a point of identification for journalism to build an "imagined community" overarching the differentiated life-worlds of its audiences.

A somewhat similar dualism plagues the theoretical field of global communication and media. Intense globalization during the latter part of twentieth century provoked theorizations about the possible emergence of transnational or "global public spheres." This debate continues today. Some commentators emphasize the growing complexity and potential of transnational communication networks to cut across national boundaries (e.g. Castells 2008; Volkmer 2015). Others view globalization as an ideology that hides actual power relationships rather than a notion that helps us observe what is actually happening (e.g. Sparks 2007; Mann 2013). Parallel debates have appeared concerning "global journalism" (e.g. Reese 2001, 2008; Hafez 2007; McNair 2006; Berglez 2008, 2013), some arguing that it is a myth, while others suggest that journalism and media are key agents of globalization.

Analyzing a multinational sample of mainstream journalistic coverage of the IPCC AR5 presents a valuable opportunity to discuss these tensions both empirically and theoretically. In this chapter, we draw on Sheila Jasanoff's (2005) concept of civic epistemologies, which she defines as "the institutionalized practices by which members of a given society test and deploy knowledge claims used as a basis for making collective choices" (ibid.: p. 255). By extending this concept to focus specifically on media and journalism, we suggest a concept of *mediated civic epistemologies*. This conceptualization helps us look at the specific, "actually existing" public spaces constructed by nationally anchored newspapers to see how they mobilize (or do not) local social actors in making sense of the report. While it is most important to understand and appreciate the particularities of these local contexts, one practical way of identifying their differences is to compare media performance, using this as a starting point to elaborate on the contexts. We examine these national contexts comparatively, elaborating on similarities and differences to gain a better understanding of the dynamics of journalism as a potentially transnational institutional force.

This chapter follows the same analytical strategy as Chap. 3 by discussing media *attention* and the *access* of social actors (voices) to the public space opened by journalism. It illustrates the diversity of contexts that underlie the performance of the press in the 22 countries involved: economy, carbon emissions, media freedom, climate change vulnerability, public perception of climate change. Standardized indicators of these background factors are, of course, complex constructions full of political choices and compromises. Hence, it is important to bear in mind that they are forms of global governance in themselves. Furthermore, they do not causally explain the performance of journalism in any simple way. In relation to media coverage, they illustrate the *structural* conditions in which newsrooms and individual journalists act to *domesticate* the global picture.

Domestication of International/Global News

It is a well-known fact that journalists evaluate events by criteria of cultural, political or economic proximity and relevance (Galtung and Ruge 1965). This diagnosis has held its basic validity since the 1960s (Wu 2000). In the early 1990s, the notion of *domestication* was introduced to news media research to argue against the overenthusiastic belief in the news as an agent of globalization (Gurewitch et al. 1991). Elaborating on early news research criteria, this new notion emphasized that for a news item to become newsworthy (to gain attention) the support of pre-existing local cultural frames and narratives are vital. Thus, research on news domestication is not limited to which internationally circulated topics are chosen for publication. More importantly, it tackles how news events are reproduced locally, concentrating on how familiar news sources and actors translate international topics and themes into localized versions, thus shaping "global" discourses for national or local consumption (e.g. Clausen 2004; Nossek 2004).

Comparative research on coverage of the "same" news events in different locations has helped to develop a more nuanced understanding of domestication. Two recent strands of argument are important. First, news domestication is increasingly viewed not only as the reception of global themes but as a practice where local news actors are drawn—as active participants and agents—into a public debate that *reconstructs* the global discourse (see e.g. Alasuutari et al. 2013; Eide et al. 2008; Kunelius et al. 2007; Olausson 2014). This view emphasizes that global discourses exist through their local articulations. In actual situations, the elements of global and local discursive spaces mutually define and construct each other. Second, there are transnational similarities and patterns of that stretch wider between individual nations but are not globally shared. Explanations of such partially homogenizing tendencies in international news have emphasized the role of international news agencies, transnational journalistic culture, ideological affinities (e.g. market liberalism), the legacy of the Cold War and state alliances (Curran et al. 2015). Both these angles—appreciating the differences in individual nations and recognizing broader clusters of countries—are relevant if we wish to understand the domestication of the IPCC message.

Our analysis in this chapter, and in the earlier work of the *MediaClimate* network (see Eide et al. 2010; Eide and Kunelius 2012; Kunelius and Eide 2012), has sought to shed comparative light on the ways in which journalism domesticates global climate politics by opening up national spaces of interpretation (see also Schmidt et al. 2013; Schäfer et al. 2014; Painter 2010, 2013). In this chapter, we discuss these parallel local public spaces in three empirical steps. First, we outline some of the structural dependencies of journalism by situating local coverage in its national, political-economic and cultural conditions of material and symbolic power. Second, we provide comparative observations from our 22 country content analysis by sketching national media responses to the AR5 according to levels of *attention* and the *access* of different actor voices to these public spaces. This step poses the challenge of grouping or clustering countries based on their media performance. It is tempting to think that one could advance from country groups to hypotheses and assumptions concerning the amount of attention and diversity of voices in the coverage. However, the complex intersection between climate politics, media systems and other contextual factors offers no such clarity. Instead, we have chosen a reverse strategy of opening up our data. Rather than offering a comparative framework where socio-economic or political factors predetermine country groupings, our approach breaks with the simple national domestication argument. This suggests there are different kinds of transnational contexts and cultures of journalism, while allowing for a more detailed presentation of the national data. Hence, we group countries according to the media's primary response (attention) and move forward by examining what types of contextual factors can produce similar (or comparable) amounts of attention. Our final step is to tentatively identify four ideal types of domestication logics or, types of *mediated civic epistemologies,* in climate politics, using them to characterize the dominant features of coverage in some countries. This theoretical step, in turn, raises new questions regarding the variety of assumptions about climate science and public debate that underlie global climate communication.

STRUCTURAL CONSTRAINTS AND COMPARISONS

To understand the diversity of climate journalism and its performance in different parts of the world, one must begin by situating local journalistic fields and institutions into the conditions in which they operate. There is no single, authoritative way of doing this, as measurements of such constraints are in many ways contested constructions. Should one use GDP per capita, for instance, for measuring the general "level" of economic development? Which media freedom index is best? Is it more relevant to look at carbon emissions per capita or to look at personal, diversified carbon footprints, based on assessments of consumption and life style? Which comparative public opinion poll is best? Such indicators are intertwined with the mobilization of the very power resources that they measure. However, they are not indicators that would be completely reduced to power relations, and thus, the ambiguity of this terrain is not a reason to ignore them nor importance of structural social constraints. Thus, from the multitude of national indicators that are relevant in global climate politics, we have chosen to collect a contextual data set that consists of the following (Table 4.1).

We use two indicators of the global carbon *economy* to situate nations. *Gross domestic product per capita* allows for comparison of the placement of a nation in the economic ranking of the world through the category of an individual person. It represents, in a rough sense, what is at stake in the climate economic future for citizens of different nations—if we assume people are narrowly rationalized economic agents. Bluntly put, it shows how much people stand to lose if the current carbon economic world order changes dramatically. What GDP *per capita* does not show, of course, is the relative economic power of *nations* as bargaining actors in the international system. Obviously, China has more leverage in global climate politics than its per capita GDP suggests.

Table 4.1   Background indicators for the 22 studied countries

	GDP[a]	CO2[b]	MFI[c]	Web[d]	HDI[e]	CRI[f]	VAI[g]	PA[h]	PRP[i]
Australia	64.43	16.92	17	83	0.93	52.8	77	98	81.9
Bangladesh	1.03	0.34	43	6.5	0.56	20.8	40.4	73.2	98.8
Brazil	11.89	2.56	31.9	51.6	0.74	85.7	54.1	77	96.4
Canada	52.39	15.67	11	85.8	0.9	95.5	76	96.2	80.7
Chile	15.69	5.47	23	66.5	0.82	101.3	66.3	78.1	96.6
China	6.96	7.42	73.6	45.8	0.72	44.7	58.6	75.6	60.4
Egypt	3.2	2.6	50.2	49.6	0.68	116.3	49.9	74.5	87.1
Ethiopia	0.52	0.08	41.8	1.9	0.44	84.2	38.3	n/a	n/a
Finland	49.21	10.57	7.5	91.5	0.88	151	80	97.5	63.4
France	44.1	5.73	21.2	81.9	0.88	43.7	73.4	94.9	82.7
Germany	46.2	10.21	11.5	84	0.91	42.7	78.2	96.1	66.8
Indonesia	3.68	1.95	40.8	15.8	0.68	72	48.4	68.9	87.7
Israel	36.07	9.02	32.1	70.8	0.89	114.3	64.6	86.4	74.6
Japan	38.63	10.7	27	86.2	0.89	91.7	73.3	99.1	81.3
Norway	77.79	8.5	7.8	95.1	0.94	131.8	81.1	97.1	64.1
Poland	13.82	8.47	12.7	62.8	0.83	65	69	84.3	70.2
Russia	14.47	12.63	45	61.4	0.78	43.3	65.9	83.9	56
S. Africa	6.89	6.25	22.1	48.9	0.66	83.8	53.9	66.2	69.9
Sweden	60.09	5.04	9.5	94.8	0.9	128.3	79.9	96.2	67
Uganda	0.69	0.05	31.7	16.2	0.48	84.8	38.3	68.3	86.1
UK	41.82	7.53	20	89.8	0.89	67	78	97.3	75.3
USA	52.94	16.55	24.4	84.2	0.91	44.7	76.5	98.4	79.7

[a]GDP per capita in thousands of USD (International Monetary Fund, 2013; Statistics Norway, 2013)

[b]Carbon emissions per capita in tonnes (EDGAR 2013)

[c]Media freedom index where the lowest score indicates more freedom (Reporters Without Borders 2015)

[d]Percentage of individuals using the Internet (ITU 2013)

[e]Human development index (HDI) estimates life expectancy and living standards, where the highest score represents better results (UN Development Programme 2013)

[f]Climate risk index for 1994–2013 where the lowest score means the highest risk (Germanwatch 2015)

[g]Vulnerability and adaptation index (ND-GAIN 2013), higher score indicates better preparedness to resist threats of climate change

[h]Public climate change awareness by percentage (Lee et al. 2015, data from Gallup World Poll 2007, 2008)

[i]Public climate change risk perception (Lee et al. 2015, data from Gallup World Poll 2007, 2008). Percentage of those indicating "serious" personal risk as a share of those deemed "aware"

Carbon emissions per capita represents national emissions, again, through the category of an individual person. What it does not capture is the amount of carbon emissions built into consumer-intensive individual lifestyles in some parts of the world; particularly those that outsource the emissions produced by their consumer products to other, differently developed societies with varying degrees of social stratification. For example, the carbon footprint of a frequent traveler from the EU (say Finland) is larger than the emissions per capita data suggests because it does not account for emissions outside the country. Hence, choosing to incorporate the "individual" in such figures does not come without consequence. Despite these statistical doubts and biases, we think that rhetorically journalism imagines and constructs its audience as citizens or individuals. Therefore, the per capita figures discussed here are at least weakly defendable as a condensed, symbolic representation of such factors.

To reflect on the adaptation, risk and vulnerability dimensions of national climate politics, we refer to three other indicators. The United Nations *HDI* (UN Development Programme 2013) combines general economic, educational, democratic and other factors that are crucial for the ability of nations to adapt to the foreseeable socio-economic consequences of climate change. The *HDI* provides a broader view of societies and their quality of life, but the measure produced correlates strongly with GDP. For a more direct indication of the risks related to climate change, we refer to the *Climate Risk Index* (Germanwatch 2015). This measure takes human and economic losses caused by extreme weather events into account. In contrast, the *Vulnerability and Adaptation Index* (ND-GAIN 2013) does not measure actual consequences but reflects the ability of people to cope with the negative impacts of climate change. Together, these three indicators (development, risk, vulnerability) highlight the *concrete* imbalance and inequality of global development, concerns that are not built into GDP and current carbon emission estimates. At the same time, *awareness* of such imbalances is a factor that shapes global climate discussions in the Global North and the Global South.

Two indicators are used to mark the general media and political conditions in which journalism operates. The *World Press Freedom* index (Reporters Without Borders 2015) reflects the legal, political and economic constraints on journalism. It incorporates several crucial aspects of the media–democracy axis that are fundamental to journalism's ability to participate in and facilitate a diverse and viable public debate. As a more general background indicator of the contemporary media environment, we refer to *Internet Penetration* rates (ITU 2013). Together, these indicators help us situate journalism in relation to the local power elite (media freedom) and offer a rough indicator of how the general population is linked as an audience to information flows.

Finally, mainstream journalism is situated in relation to representations of local public opinion. With reservations about how to measure "public opinion" withstanding, assumptions about the level of awareness and seriousness of climate change among the general population influences the media agenda. For a comparative perspective, we draw on a recent study of the public perception of climate change (Lee et al. 2015) that relies on data from the *Gallup World Survey* for 2007–2008. Although the sample could be more recent, we chose to work with this date because of its wide global reach and due to the fact the survey offers a unique global comparison of *public awareness* and *personal risk perceptions*. The authors argue that "…widely used national indicators of sustainability (for example, HDI and GDP) are poor predictors of the correlative structure of climate change awareness and risk perception globally" (ibid.: p. 6). However, they point to a wide array of individual level explanatory factors across the globe. These range from level of education and access to communication to how much direct experience of exceptionally high temperatures people report.

There are two reasons why these indicators provide a necessary background for an analysis of the global–local nexus and the domestication practices in media coverage of the IPCC AR5. First, and almost self-evidently, despite the inherent political issues and methodological problems of measurement, the indicators refer to real material and structural conditions that help to understand the nature of climate coverage in different countries and contexts. For instance, the low amount of attention in our Russian or Chinese samples can be understood against the combination of state media control *and* the respective energy-power interests. Low coverage in some developing countries may be interpreted as resulting from the lack of newsroom resources, the overall economic situation *and* lower public climate awareness, which, in turn, reflects the pressing importance of other local issues and problems. A second reason to pay attention to these indicators is that, arguably, these indicators represent many of the implications and assumptions that are strongly present in the ways journalists imagine the global scene and situate their own nation within it. Material constraints and pressures are always symbolically mediated when they enter the fields of public discourse, media and politics. The indicators, then, help account for the symbolical and material conditions under which mainstream newspapers open up national spaces of interpretation—spaces that are ultimately the building blocks of the global climate discourse.

DOMESTIC ATTENTION POLITICS AND ACCESS OF VOICES

The relationship between country indicators and differences in media attention evades simple explanations but has helped to point to some important links. For instance, Schmidt et al. (2013) have shown that countries with significant Kyoto commitments tend to show higher media attention to climate and that the carbon intensity of a given national economy or an extensive role in fossil fuel exports can be linked to higher media attention (See also Schäfer et al. 2014; Djerf-Pierre 2012).

In our IPCC material, correlations between the total attention score (see Sidebar 1.1 and Chap. 3 for explanations on how the scores were measured) and different country indicators (Table 4.1) suggest the following overall result. GDP and HDI generally correlate positively with the amount of attention in the total sample. Public interest and the strong potential for climate science to attract public attention, then, coincide with the general level of development. High carbon emissions per capita is generally intertwined with this index of "development," however in *some* countries it can also translate into lower media attention. This seems to be the case, particularly, in countries outside the 1997 Kyoto agreement (i.e. countries with insignificant or low commitments). In our sample, the notable examples are Russia and China, where low media interest coincides with high national stakes in climate and energy politics, low media freedom and strong governmental control of the mainstream media. Clearly, in these cases journalism is unable to use the global scientific authority of the IPCC to raise the importance of the issue in the national news agenda, let alone to bring about a diverse local political debate.

Climate vulnerability and adaptation scores are not strongly systematically linked to IPCC media attention. However, in some nations a high vulnerability ranking notably coincides with low media attention. Thus, even if future risks and adaptation problems are known to be high, this does not result in media sensitivity to climate science and the IPCC message. In our sample, this trend is common for countries representing a group of emerging economies (China, Russia, South Africa). This underscores a structural defect in journalism's systemic ability to draw public attention to future hazards faced by the populations it aims to serve. Signs of a contrary relationship between vulnerability and attention occur in the cases of Australia and the USA where, relative to other countries in the sample, the IPCC report and climate science attracted more media attention than "vulnerability" levels would suggest. In such cases, other factors explain the higher attention levels better. Overall, one can say that vulnerability does not translate effectively into a higher sense of risk and thus media attention. In the case of developing countries this can be put more sharply: low media attention and low resources for climate journalism are a key part of the vulnerability itself.

We can also compare the attention scores between the total sample (all stories that mention climate change during the sampling period, N = 1615) and the core sample (stories where AR5 is the main topic, N = 551). If we consider the total sample of attention as an indicator of higher "background" readiness to write about climate (i.e. higher basic sensitivity in local journalism for using "climate change" as part of their news narrative), we can see that this also influences the amount of attention paid to the report (e.g. the UK, Brazil, France or Sweden). In our sample, Australia, the USA and Finland, with Japan and Canada to a lesser extent, stand out with their higher level of "background" reporting on climate during the sampling periods—rather than coverage driven by an explicit focus on AR5. In Australia and the USA, this can be explained by the polarized local politicization of climate change. This increases the general coverage since having high political stakes in climate change action encourages stakeholders to seek out media access and exposure. Yet, at the same time, higher public tension made the AR5 more "controversial," leaving the IPCC with a less authoritative role as a source of knowledge and topic of attention.

A key argument in the news domestication debate has concerned the role of local sources as coproducers and translators of global news into local frames. Figure 4.1 plots the 22 countries, based on their AR5 sample, against two dimensions: the "strength" of domestication (percent share of domestic sources of all voices) to the relative media attention.

The lower-left corner of the figure shows countries where the mobilization of domestic news actors as voices in the coverage was minimal. In these countries, where media attention to the AR5 was also very low, foreign and transnational voices clearly dominated the news. Here the developing countries in our sample stand out: Ethiopia, Egypt, Bangladesh, Indonesia and Uganda. This lack of attention is not the responsibility of journalists alone. It is also reflective of a broader problem news coproduction face in local contexts where political or scientific elites are rarely engaged in the IPCC launch process. The local public space of interpretation in these countries, then, is weak; both in terms of volume and in terms of reliance on foreign or transnational voices. In the upper-right corner, we find countries with a high number of domestic voices mobilized. In Australia, Canada, Finland, France,

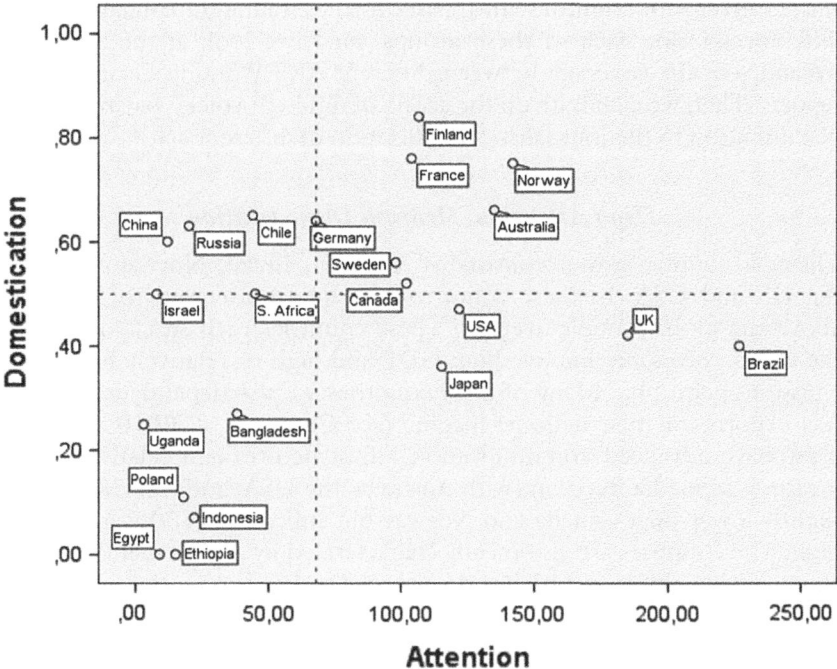

Fig. 4.1 Domestication versus attention (AR5 sample, 780 stories, 1332 voices)

Germany and Norway, more than half of the voices were local actors. This supported a higher amount of media attention. In Japan, the USA, UK and Brazil the pattern was similar but slightly less domesticated. The upper-left corner, on the other hand, clusters Chile, China, Russia, Israel and South Africa; nations where low media attention led to weaker public spaces, albeit still rather strongly dominated by domestic actors. In these countries, the role of *foreign* voices (and not transnational) was also more pronounced.

Attention and Access: Local Dynamics

There was no simple way to cluster all 22 countries due to the multiple combinations of background variables and the diverse national news agendas during the AR5 launch periods (see Sidebar 1.1 and Chap. 3). However, in order to bring more nuances to this overall picture, we have combined the country factors data with the attention scores to form four smaller groups of countries. In this section, we highlight similarities and differences inside each of these groups. First, we look at the attention dynamics of the coverage between different AR5 Working Group (WG) reports. Then we elaborate on the access of different voices, paying particular attention to the journalistic mobilization of different actor categories.

High Attention, Medium Domestication

The first country group consists of Australia, Brazil, Norway, Canada, the UK and USA. In these countries, the media attention to AR5 was above the global sample average. These countries each situated high in the carbon economy figures: high GDP and high or relatively high CO^2 emissions per capita. Many of these countries are also dependent on fossil fuel exports for their national income (see Chasek et al. 2010: p. 182). They have advanced communication infrastructures and relatively high measures of media freedom (with Australia, the USA, and the UK scoring slightly lower than Canada and Norway but still well). HDI is extremely high. The countries are also mostly characterized by a combination of low climate vulnerability and high potential readiness for adaptation measures. In the global comparison, public awareness of climate change is high in these countries (Lee et al. 2015) while personal perceptions of climate risk are not. This may reflect the low vulnerability indices in these countries. In most of the countries in this group, then, the potential political-economic interests and stakes for climate change action are high, people also are aware of climate change—but they perceive its threats as less pressing. In our sample, Brazil belongs to this high media attention group.[1] However, Brazil is an outlier when it comes to other country indices: it scores lower than the others on GDP, emissions, media environment and somewhat lower on HDI. Like others in the group, it has a high public awareness score, but this awareness is linked to even higher perceptions of risk among the public.

Looking at the AR5 publications and their attention dynamics inside this group (Fig. 4.2), Norway stands out with a more steady level of attention throughout the coverage. Otherwise, the AR5 sample follows the

---

[1] This can be explained, in part, by the fact that climate change was a key theme in the campaign of one 2014 presidential candidate, similar to the 2010 elections when three of the main Brazilian candidates visited COP15 (see d'Essen 2010: p. 83).

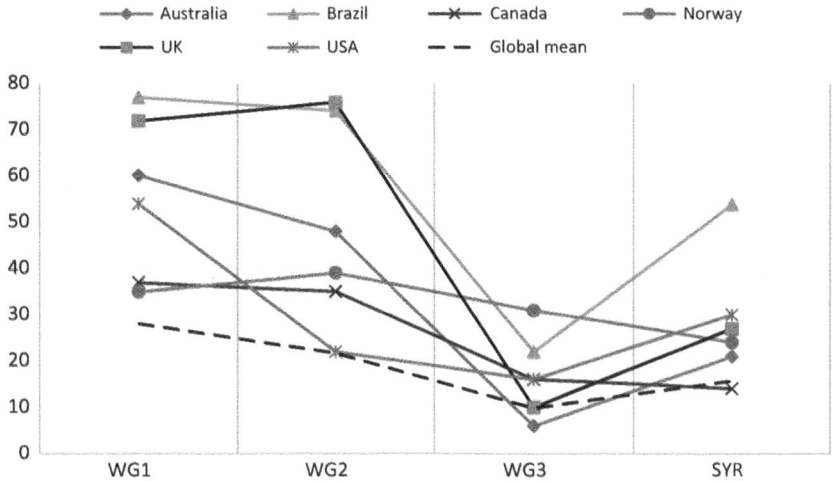

Fig. 4.2 Attention dynamics in country group 1 (AR5 sample)

same general, global logic (Chap. 3) with WGI generating high attention levels followed by a steep decline for WGIII. As noted above, media attention in Australia and the USA (and Canada to a lesser extent) was more explicitly focused on the AR5 than on climate generally. This may reflect the intensively politicized issue context. On the contrary, the coverage in Brazil, Norway and the UK was more driven by a specific focus on the IPCC results (affording more authority and centrality to the IPCC). The differences within these clusters illuminate broader differences in their political and journalistic cultures. In Australia, the USA and Canada, to a lesser extent, political systems and media realities rely on a *civic epistemology* where truth is more a matter of public confrontation, whereas in the UK and Norway the empirical, scientific reasoning of the IPCC holds more authority (Jasanoff 2011: p. 135).

Casting an eye on *voices* in the news stories points to particular similarities and differences. Table 4.2 shows the main categories of voices in each of the six countries, with separate rows highlighting the share of domestic voices (i.e. Norwegian voices in Norwegian coverage, etc.) in the overall coverage. Not surprisingly, scientists play a prominent role across the whole country group. The numbers also confirm the politicized logic of the USA and Australian coverage, and show a relatively high number

Table 4.2 Country group 1 voices in the AR5 sample (percent from *N* of voices)

	Australia (N = 125)	Brazil (N = 167)	Canada (N = 103)	Norway (N = 122)	UK (N = 173)	USA (N = 88)
Transnational politics	8	6	8.7	3.3	13.3	12.5
National politics	23.2	7.2	9.7	25.4	19.1	23.9
domestic	17.6	3.0	5.8	23.8	12.1	18.2
Civil society	14.4	15.6	26.2	23	28.9	14.8
domestic	12.8	5.4	24.3	21.3	13.9	5.7
Business	3.2	3	1	4.9	4	3.4
domestic	2.4	1.2		3.3	1.7	1.1
Science	43.2	64.7	50.5	42.6	30.1	33
domestic	32.8	29.9	22.3	27	13.9	21.6
Media	8	3.6	3.9		2.9	12.5
Other				0.8	1.7	

of domestic politicians combined with a high proportion of domestic scientists.

What distinguishes Canada, Norway and the UK from the USA and Australia seems to be a different *kind of politicization* that involves civil society actors more. In Norway and Canada, civil society actors are more domestic. In the UK, the global orientation of *The Guardian* shows clearly: all foreign NGO voices in the UK sample were from *The Guardian*, which quoted civil society actors four times more often than *The Telegraph*.

Inside this group of countries, then, we can detect two kinds of high attention domestication logics. On the one hand, there are the polarized climate political cultures of the USA and Australia, where national politics is rather dominant, scientific voices tend to come from networks of domestic expertise, and civil society voices are noteworthy but less important than political ones. On the other hand, there are contexts in which domestication is driven more by civil society (Canada, Norway and the UK) and where journalism turns more actively to the NGOs. Norway, in particular, shows a profile that at least allows for a vibrant *possibility* of interaction between domestic politics, domestic civil society and domestic science (together, these three categories represent 72 percent of the actors). In Norway, political and journalistic traditions generally support this diversity. However, in climate journalism it can be heightened by a conflicting national self-identification as an environment-friendly oil exporter, the "tainted hero" of global climate politics (Eide and Ytterstad 2011).

Canadian journalism shows signs of domestic political withdrawal from the issue, allowing more room for (transnational) scientists.

Brazil offers a different example of how high media attention may be *without* a strong, visible link to domestic actor networks in journalism. From the actors' perspective, the Brazilian coverage was rather depoliticized. The weakness of domestic civil society voices and the withdrawal of domestic politicians meant the field was largely left to scientists.

Medium Attention, High Domestication

The second group of countries consists of Finland, France, Germany, Japan and Sweden. Here, media attention is still relatively high (hovering around or above the global average for the AR5 sample). In terms of structural background indicators, the group is relatively homogeneous: high GDP, relatively high emissions per capita (Finland, Germany and Japan) with some diversity (smaller emissions in France and Sweden), high or relatively high media scores (with Japan and France lagging a bit behind others) and very high HDI rankings. While the countries in the previous group had vital national interests in the carbon economy, countries in this group have few fossil fuel resources and are relatively dependent on imported energy (see Chasek et al. 2010: p. 182). In terms of climate vulnerability and adaptation ability, these countries all belong to a privileged low vulnerability grouping with high capacity to tackle the consequences of climate change. Public awareness of climate change tends to be very high, though people's assessment of personal risk related to climate change is moderate.

The attention dynamics between different AR5 WGs (Fig. 4.3) shows Finland and Sweden following the general global trajectory. Germany shows more sustained coverage throughout while, in France, after an initially high attention peak to WGI, the interest level drops and remains low. In Japan, interestingly, the attention level has a slow declining tendency throughout. Beyond the figure, though, Japan is a notable exception in the general climate coverage (the total sample) where the attention level rises clearly toward the end. In France, Germany and Sweden, the coverage is more driven by the core IPCC stories.

In the European countries of this cluster, domestic politicians play a consistent role in the coverage, but, unlike other countries, Sweden includes considerably more transnational voices (Table 4.3). In Sweden and Finland, civil society voices are an important part of the coverage, resembling Norway (see above), but with the domestic civic actors playing a smaller role in

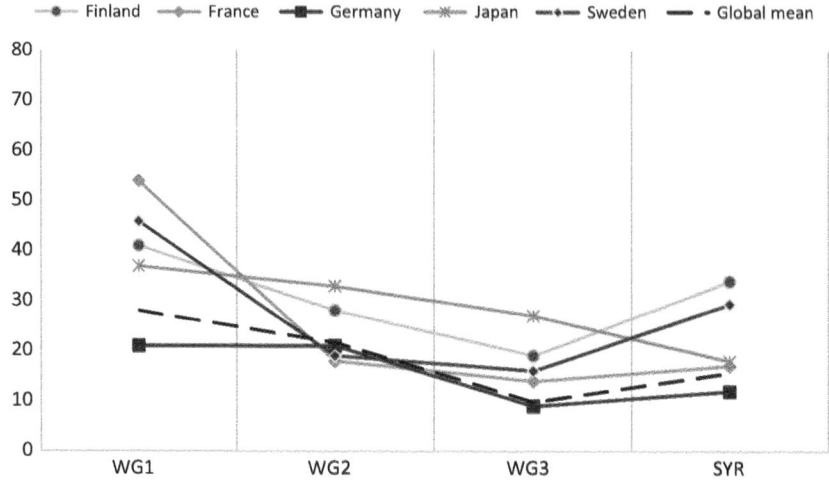

Fig. 4.3   Attention dynamics in country group 2 (AR5 sample)

Table 4.3   Country group 2 voices in the AR5 sample (percent from *N* of voices)

	Finland (N = 63)	France (N = 76)	Germany (N = 44)	Japan (N = 73)	Sweden (N = 66)
Transnational politics	7.9	5.3	11.4	6.8	18.2
National politics	12.7	13.2	22.7	13.7	19.7
domestic	12.7	11.8	11.4	1.4	15.2
Civil society	22.2	2.6	4.5	9.6	19.7
domestic	22.2	1.3	4.5	6.8	6.1
Business	11.1	5.3		2.7	
domestic	9.5	5.3		2.7	
Science	44.4	72.4	61.4	64.4	42.4
domestic	39.7	57.9	47.7	24.7	34.8
Media	1.6	1.3		2.7	

Sweden. This corresponds to our previous findings that show exceptionally strong positions of civil society groups in Nordic countries (Kunelius and Eide 2012: p. 282). Broadly speaking, in the European context, this demonstrates the tendencies of "corporatist" media traditions that encourage involvement from various social groups (Hallin and Mancini 2004). France and Germany represent a distinctly different case, with a smaller proportion of civil society voices in the coverage. In Japan, their role is even smaller. The voices of domestic scientists dominate in all of the European countries.

Overall, the European countries examined here reveal two kinds of domestication cultures, one with a higher visible role for civil society networks (Finland and Sweden) and the other with more dominant scientific actors (France and Germany). Japan shows some characteristics similar to France (high reliance on science), but it also represents a different a kind of domestication logic where domestic politicians seem extremely withdrawn from the actual IPCC debate (Chap. 8).

Low Coverage, High Domestication

The third group of countries consists of China, Poland, Russia and South Africa, where media attention to the AR5 was very low. The GDP per capita in these countries is lower than in the countries of the previous groupings. However, their emissions per capita are relatively high (the same level as in the previous group). Media freedom indicators here vary considerably: China and Russia score very low, while Poland and South Africa score close to European levels. In terms of HDI, Poland is closest to the European countries while the others score lower. None of these countries belong to the most immediately climate vulnerable locations in the world. Although their readiness to tackle climate issues is not as high as the previous groupings, the relation between their specific challenges and adaptation potential is comparatively good. All these countries demonstrate lower levels of public awareness of climate change combined with rather low personal perceptions of risk. Roughly put, then, there is less immediate public opinion pressure for media to pay attention to climate issues. At the same time, there are strong political and economic factors that support the use or export of fossil fuel energy. In addition, there are notable restrictions of media freedom at play in some countries.

The number of news articles in these countries is mostly too low to make reasonable quantitative claims about the coverage, but this, of course, is a strong result in itself. The main conclusion concerning these countries then is that the IPCC AR5 had very little effect on activating a *public* space for interpretations related to the climate (the AR5 sample). To put this differently: the IPCC's global scientific effort could not support local journalism—or other local public actors—in countries opposed to having such a space. The coverage in China reflects a disciplined state-science communication effort with virtually no explicit attempt to engage other actors. It offers an example of a coverage that restricts the issue to an official policy question (both domestically and internationally). In Poland, the selected newspapers invested very little editorial energy on the issue, citing only transnational politics and a few domestic scientists in a handful of stories, and showing a markedly different kind of attitude than their European neighbors. In Russia, media attention to the IPCC was also very low. Civil society actors and scientists play a visible role, but that role either verifies and confirms official policy or avoids political questions. South Africa stands out with a slightly larger sample and more diversity of voices, though mostly in relation to WGI. For reasons of consistency, we present the similar coverage data from these countries as well (Fig. 4.4 and Table 4.4), despite the limited number of cases. (Chap. 7 for further discussion of key countries in this grouping.)

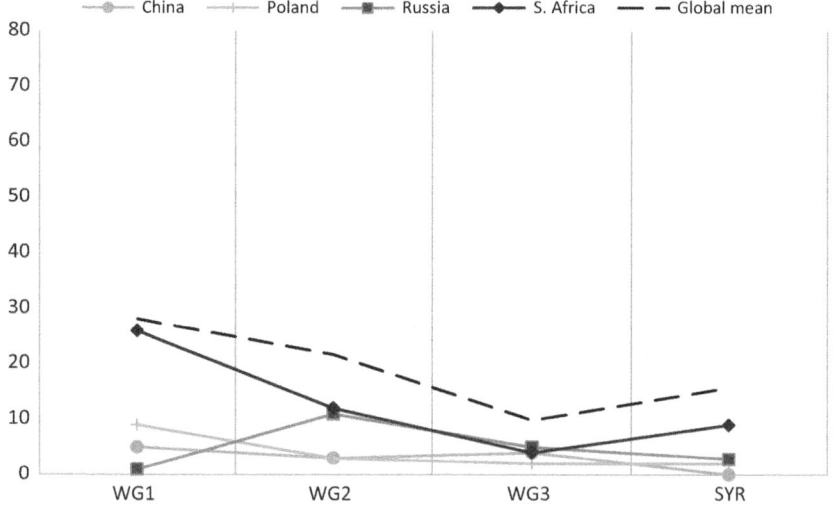

Fig. 4.4 Attention dynamics in country group 3 (AR5 sample)

Table 4.4 Country group 3 voices in the AR5 sample (percent from N of voices)

	China (N = 5)	Poland (N = 9)	Russia (N = 19)	S. Africa (N = 52)
Transnational politics		44.4	5.3	9.6
National politics	20		5.3	21.2
domestic				11.5
Civil society			26.3	15.4
domestic			26.3	7.7
Business				5.8
domestic				
Science	80	55.6	63.2	46.2
domestic	60	11.1	36.8	30.8
Media				1.9

Low Attention, Low Domestication

The fourth group of countries consists of Bangladesh, Chile, Egypt, Ethiopia, Indonesia, Israel and Uganda. As in the previous group, media attention to AR5 was very low. In terms of GDP per capita and carbon emissions, Chile and Israel are on the high end of this group; they also score well on HDI. They are also relatively advanced in terms of media environment. Bangladesh, Egypt, Ethiopia and Uganda clearly belong to the developing world, both in terms of GDP, HDI and media indicators, as well as in their very low emission figures. As for climate vulnerability, Chile and Israel are more privileged than the rest of the group that are quite susceptible to future climate hazards and poorly prepared for adaptation. General public awareness about climate change is consistently low throughout this country group. However, the perceptions of personal risks related to climate change are rather high among those who are aware. Thus, the group represents a counter-image of the first two, and a different kind of case than the previous lower attention group.

Again, the most dominant finding here is the low amount of coverage (Chap. 10 for an elaborated discussion on the reasons for this). What differentiates this group in the overall global comparison of attention dynamics is that in some countries the Synthesis Report gained more attention (except Chile and Ethiopia, see Fig. 4.5). This is understandable when one looks at the way the IPCC framed its communications: the press message of the Synthesis Report was clearly more geared toward global politics

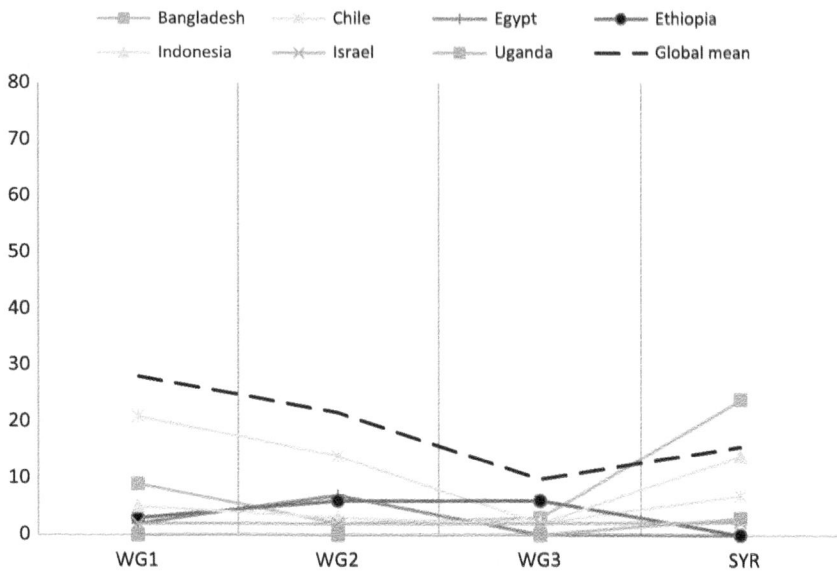

Fig. 4.5 Attention dynamics in country group 4 (AR5 sample)

Table 4.5 Country group 4 voices in the AR5 sample (percent from *N* of voices)

	Bangladesh (N = 30)	Chile (N = 48)	Egypt (N = 10)	Ethiopia (N = 23)	Indonesia (N = 30)	Israel (N = 2)	Uganda (N = 4)
Transnational politics	33.3	4.2	60	26.1	56.7		75
National politics	23.3	10.4	10	21.7	36.7		
domestic		10.4			6.7		
Civil society	13.3	8.3		13	3.3		
domestic	3.3	4.2					
Business domestic			10				
Science	26.7	77.1	20	39.1	3.3	100	25
domestic	23.3	50				50	25

and action (Chap. 3), with more UN representation and specific development politics undertones. This global political message—rather than a focus on specialized reports on mitigation challenges or adaptation—seems to have resonated with the mainstream press conditions in this group. A partial explanation for this is the overreliance on syndicated content from international press agencies due to a lack of domestic journalistic resources. Voices in the small number of stories in this country group show the importance of transnational political actors in the interpretive space of these countries. The local elites are not actively involved in this domestication process. With the exception of Chile, we also see the limited role played by scientists in general and domestic scientists in particular (Table 4.5).

MEDIATED CIVIC EPISTEMOLOGIES: FOUR PRELIMINARY IDEAL TYPES

It is important to bear in mind that the category of "voices" in the analysis only refers to social actors that journalists explicitly mentioned and quoted in the coverage. Thus, our comparative analysis only gives a partial view of the domestication (through interactions between journalism and other local actors) in local spaces of interpretation. The results do not say anything conclusive about the structures of influence nor about mechanisms of deciding whether to devote attention to the IPCC or not. The results also cannot fully account for the dialogic diversity of journalists writing with their own distinct voices: numbers are not voices. Instead, the results highlight the kinds of actors and relationships that journalists make visible and consider important enough to be played out in the public. With these limitations aside, focusing on the single "event" of the AR5 launch and the different ways journalism in 22 countries created public interpretive spaces to discuss the IPCC findings enables us to catch a glimpse of "actually existing global journalism." We can draw some tentative conclusions by situating these findings within the ongoing discussion of the role news media play as institutions of domestication.

Any discussion concerning the logics of domestication in the field of global climate politics must consider the *structures of global communication flows*. The ability of the mainstream press to draw attention to a global, complicated source like the AR5 is heavily constrained by circumstances of economic resources and communication landscapes. The best predictors of media attention in our sample are the development indexes (HDI or GDP). Although they do not predict the intentions and motivations of journalists, they do capture the limits of their practical resources. These constraints largely follow the well-known contours of global economic and development inequalities. They seem to leave the most vulnerable people with less information about the forthcoming challenges and detach the majority of politicians in the most vulnerable locations from their local publics (see Saleh 2012).

We recognize that our sample does not accurately reflect all the diversity of communication landscapes in different locations. Indeed, in some countries, the mainstream press has little to do with the life of ordinary people, whereas elsewhere new means and networks of climate communication perhaps mediate climate science more effectively to particular audiences. This is hopefully the case somewhere and for some actors. Yet, this look at the mainstream press underlines some serious "flaws" in its performance, especially in terms of building and supporting local spaces of debate and interpretation. If the task of the mass media is to help open interpretive spaces where global (IPCC) knowledge is transformed into locally meaningful dialogues, the overall results here are not encouraging. Even if we bracket out the goal of *actually informing* the public about climate change and think of the press as representing an *informed image* of the public that pressures local decision-makers, the first overall lesson of this comparative overview is not uplifting. In addition to the structural deficiencies mentioned above, the findings suggest that in many locations we need more nationally grounded domestication efforts from journalism *before* any effective transnational or global publicity can emerge. An advancement of "global journalism" that would be locally meaningful (see Berglez 2013) would also demand changes in the institutional structures (politics, science, civic activism) in which journalism is locally embedded.

The analysis, then, shows the contours of global climate politics and the way journalism, particularly the elite-oriented mainstream press, is controlled and structured by it. The most obvious reflections of this can be seen in the low attention and high domestication group of countries; particularly in Russia and China, where national energy goals and global power politics influence the media environment. In such contexts, the general lack of communication resources—or the lack of supportive expert infrastructure for writing about climate science—are not enough to explain (or justify) the politics of low attention. Rather, we see this as a link between power, energy choices, politics and the ways in which journalism (mis)represents the public. It is important to note that the same forces—the organization of societies and their power structures around long-term energy choices—are visible in other country clusters as well (Chap. 7).

It is true that local, national fields of power are crucial for particular kinds of domestication. Our analysis aims to conceptually enrich understandings of the role global climate journalism plays in that process. For forthcoming research, we conclude by sketching out four ideal types of domestication logics. As an extension of Jasanoff's (2005, 2011) conception of civic epistemologies, the analysis above suggests identifying four kinds of *mediated civic epistemologies*. While individual countries in our sample only partly suit these ideal types, this tentative theoretical framework helps to identify the contextual elements of domestication that are emphasized in different variations. They can be roughly mapped out using two conceptual dimensions, the *epistemic* focusing on the role of science (authority vs. contested/politicized) and the *civic* focusing on the cultural role of publicity (strategic vs. deliberative/reasonable) (Fig. 4.6).

A *statist*-mediated civic epistemology (or *nationalistic culture of political realism*) demands a powerful narrative about the state as a naturalized actor in international climate politics. This position downplays political identity differences inside the national public domain. It relies on the authority of *national* science as the mediator of IPCC results to the local context. The truth in such a context is something that the state eventually represents and guarantees, and thus the authority of science requires state sponsorship. In this type, the state also controls the media environment and civic participation. This nationalism works to exclude transnational networks of science or civic activism. This is explicitly the dominant culture

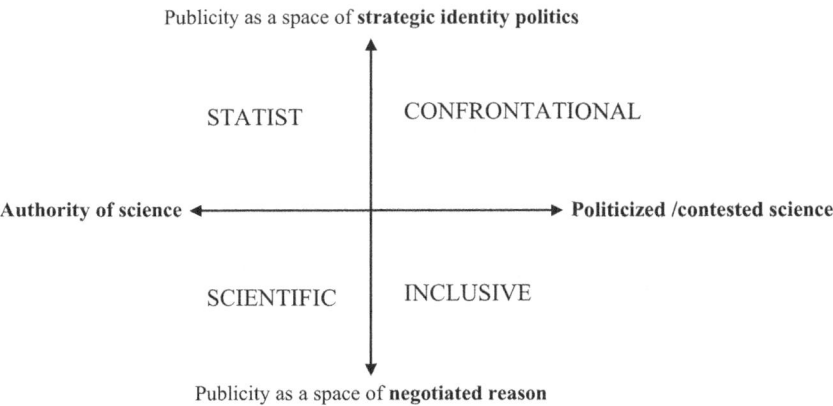

Fig. 4.6 Mediated civic epistemologies of domestication

in some emerging economies (China, Russia). Science communication in such cultures largely takes place under the patronage of national political actors. While we can link this ideal type to particular country examples, it is important to note that its realistic discourse of power is a strong element of the domestication discourses in other places as well—although science would not be as explicitly an ally of state interests.

A *confrontational culture* of climate politicization emphasizes differences in political identity within national domains of domestication. This ideal type puts the authority of science to the test of free public debate, arguing that the "truth" can be found in the encounter of opposing views. In order to function, it demands a liberal market regime of the media. Seeing the public sphere as a strategic battleground for meanings and truths, this culture nurtures media management skills on all sides of climate politics. The heightened stakes in the media field draw in media as well as actors in climate politics. Such a culture can be seen as potentially encouraging transnational communication networks. Identity-based political action opens collaborations across national borders (driven by business interests, ideology and civic action interests) and across institutional boundaries (turning scientists into public activists, forging alliances between journalism and activism, for instance). Signs of this culture are visible in some liberal democracies, particularly in countries where the fossil fuel industry is a powerful actor (USA, Australia).

An *inclusive culture* of climate politicization recognizes political differences within national domains but this ideal type ultimately assumes that differences can be rationalized in the political process (first within the nation-state and by analogy within the international community). In this rationalizing process, science plays the specialized roles of authority, guardian of knowledge and provider of expert advice; thus as something that can still set limits for what can be accepted as rational arguments. This culture demands a free media system while recognizing the need for public (service) communication that cannot be met by market mechanisms alone. Seeing the public sphere as a site of reasonable negotiation (rather than *merely* a battlefield), it emphasizes the role and duty of journalists to create diverse and inclusive debates, drawing different (e.g. civil society) voices into the space of interpretation. This culture of mediated civic epistemology seems to thrive in some of the "democratic corporatist" countries of Northern Europe (Norway, Finland, Sweden), but some of its characteristics are also part of a broader professional logic of journalism.

Finally, in a somewhat more diffuse form, there we can point to a fourth shape: a *culture of scientific universalism*. According to this kind of domestication logic, science, at least formally, enjoys universal and depoliticized authority. This culture resembles the previous one by lending science a status of the highest authority, emphasizing the power of scientific knowledge. The difference is that this epistemology downplays the role of political actors who remain in the background and more distanced from the debate. In exercising and reproducing this culture, journalism draws on the transnational, universal authority of science, sometimes combining this with the authority of transnational political (and civic) actors. The emphasis on science results in lower visibility for local political actors and their interests. In our sample this profile is, somewhat differently, accentuated in countries like France (strong voice of domestic science), Japan (more transnational science, high withdrawal of politics), Brazil and Chile. While such a tendency is perhaps more typical for some countries in our sample, it is important to underscore the strong overall role of scientific voices in the coverage.

Demonstrating how these four subcultures of mediated civic epistemologies are articulated in particular locations captures some of the forces at play in the local dynamics of domesticating the IPCC AR5. It also helps identify the local structural and cultural landscape of power in which journalists, scientists and civic actors are embedded when they attempt to translate the IPCC message into public discussion. Hopefully, the two dimensions—the *epistemic* (articulated by assumptions of scientific authority) and the *civic* (articulated by assumptions about the political meaning of public spheres)—help us to develop more nuanced and self-reflective analyses of the landscape of climate communication. In this chapter, we have shown how such ideal types emerge from the actually existing local conditions of globalizing climate journalism. In a sense, these ideal types also articulate the forces and demands that all journalists must struggle with or, to put it more constructively, some of the limits they need to push back. As the discussion in this chapter shows, this journalistic struggle takes place in incredibly diverse circumstances, which demonstrates the radically uneven robustness of the transnational communication infrastructure of climate politics.

BIBLIOGRAPHY

Alasuutari, P., Qadir, A., & Creutz, K. (2013). The domestication of foreign news: News stories related to the 2011 Egyptian revolution in British, Finnish and Pakistani newspapers. *Media, Culture & Society, 35*(6), 692–707.

Anderson, B. (2006). *Imagined communities: Reflections on the origin and spread of nationalism* (Rev. ed.). London: Verso.

Berglez, P. (2008). What is global journalism? *Journalism Studies, 9*(6), 845–858.

Berglez, P. (2013). *Global journalism*. New York: Peter Lang.

Castells, M. (2008). The New Public Sphere: Global Civil Society, Communication Networks, and Global Governance. *The ANNALS of the American Academy of Political and Social Science, 616*(1), 78–93.

Chasek, P. S., Downie, D. L., & Brown, J. W. (2010). *Global environmental politics* (5th ed.). Boulder: Westview Press.

Clausen, L. (2004). Localising the global "domestication" processes in international news production. *Media, Culture and Society, 26*(1), 25–44.

Curran J., Esser F., Hallin, D. C., Hayashi, K., & Lee, C.-C. (2015). International news and global integration. A five-nation reappraisal. *Journalism Studies*. doi: 10.1080/1461670X.2015.1050056. Published online 3 August 2015.

d'Essen, C. (2010). Brazil: COP15 as a political platform. In E. Eide, R. Kunelius, & V. Kumpu (Eds.), *Global climate, local journalisms: A transnational study of how media make sense of climate summits* (pp. 83–96). Bochum: Projekt Verlag.

Djerf-Pierre, M. (2012). When attention drives attention: Issue dynamics in environmental news reporting over five decades. *European Journal of Communication, 27*(3), 291–304.

EDGAR. (2013). *CO2 time series 1990–2013 per capita for world countries*. EDGAR Emission Database for Global Atmospheric Research, European Commission. Retrieved from http://edgar.jrc.ec.europa.eu/overview.php?v=CO2ts_pc1990-2013

Eide, E., Kunelius, R., & Kumpu, V. (Eds.). (2010). *Global climate, local journalisms: A transnational study of how media make sense of climate summits*. Bochum: Projekt Verlag.

Eide, E., & Kunelius, R. (Eds.). (2012). *Media Meets Climate: The Global Challenge for Journalism*. Göteborg: Nordicom.

Eide, E., Kunelius, R., & Phillips, A. (Eds.). (2008). *Transnational media events: The Mohammed cartoons and the imagined clash of civilizations*. Göteborg: Nordicom.

Eide, E., & Ytterstad, A. (2011). The tainted hero: Frames of domestication in Norwegian press representation of the Bali climate summit. *The International Journal of Press/Politics, 16*(1), 50–74.

Galtung, J., & Ruge, M. H. (1965). The structure of foreign news: The presentation of the Congo, Cuba and Cyprus crises in four Norwegian newspapers. *Journal of Peace Research, 2*(1), 64–91.

Germanwatch. (2015). Global climate risk index 2015. Released by Germanwatch. org. Retrieved from http://germanwatch.org/en/download/10333.pdf

Gurewitch, M., Levy, M., & Roeh, I. (1991). The global newsroom: Convergences and diversities in the globalization of television news. In P. Dahlgren & C. Sparks (Eds.), *Communication and citizenship: Journalism and the public sphere in the new media age* (pp. 195–215). London: Routledge.

Hafez, K. (2007). *The myth of media globalization.* Cambridge: Polity.

Hallin, D. C., & Mancini, P. (2004). *Comparing media systems: Three models of media and politics.* New York: Cambridge University Press.

International Monetary Fund. (2013). World economic outlook database. Released by IMF. Retrieved from http://www.imf.org/external/pubs/ft/weo/2015/01/weodata/index.aspx

ITU. (2013). Global ICT developments. Released by International Telecommunication Union. Retrieved from http://www.itu.int/en/ITU-D/Statistics/Pages/stat/default.aspx

Jasanoff, S. (2005). *Designs on nature: Science and democracy in Europe and the United States.* Princeton, NJ: Princeton University Press.

Jasanoff, S. (2011). Cosmopolitan knowledge: Climate science and global civic epistemology. In J. S. Dryzek, R. B. Nordgaard, & D. Schlosberg (Eds.), *The oxford handbook of climate change and society* (pp. 129–143). Oxford: Oxford University Press.

Kunelius, R., & Eide, E. (2012). Moment of hope mode of realism. On the dynamics of a transnational journalistic field during UN climate change summits. *International Journal of Communication, 6,* 266–290.

Kunelius, R., Eide, E., Hahn, O., & Schröder, R. (Eds.). (2007). *Reading the Mohammed cartoons controversy: An international analysis of press discourses on free speech and political spin.* Bochum: Projekt Verlag.

Lee, T. M., Markowitz, E. M., Howe, P. D., Ko, C.-Y., & Leiserowitz, A. A. (2015). Predictors of public climate change awareness and risk perception around the world. *Nature Climate Change, 5,* 1014–1020.

Mann, M. (2013). *The sources of social power: Volume 4, globalizations, 1945–2011.* Cambridge: Cambridge University Press.

McNair, B. (2006). *Cultural chaos: Journalism, news and power in a globalized world.* London: Routledge.

ND-GAIN. (2013). The Notre Dame global adaptation index. Retrieved from http://index.gain.org/matrix

Nerone, J. (2015). *The media and public life: A history.* Cambridge: Polity.

Nossek, H. (2004). Our news and their news: The role of national identity of the coverage of foreign news. *Journalism, 5*(3), 343–368.

Olausson, U. (2014). The diversified nature of "domesticated" news discourse: The case of climate change in national news media. *Journalism Studies, 15*(6), 711–725.

Painter, J. (2010). *Summoned by science: Reporting climate change at Copenhagen and beyond.* Oxford, UK: Reuters Institute for the Study of Journalism.

Painter, J. (2013). *Climate change in the media: Reporting risk and uncertainty.* Oxford, UK: I.B. Tauris and Reuters Institute for the Study of Journalism.

Reese, S. (2001). Understanding the global journalist: A hierarchy-of-influences approach. *Journalism Studies, 2*(2), 173–187.

Reese, S. (2008). Theorizing a globalized journalism. In M. Löffelholz & D. Weaver (Eds.), *Global journalism research: Theories, methods, findings, future* (pp. 240–252). Malden, MA: Blackwell.

Reporters Without Borders. (2015). *World press freedom index.* Released by Reporters Without Borders. Retrieved from http://index.rsf.org/#!/index-details

Saleh, I. (2012). Ups and downs from Cape to Cairo: The journalistic practice of climate change in Africa. In E. Eide & R. Kunelius (Eds.), *Media meets climate: The global challenge for journalism* (pp. 49–65). Göteborg: Nordicom.

Schäfer, M., Ivanova, A., & Schmidt, A. (2014). What drives media attention for climate change? Explaining issue attention in Australian, German and Indian print media from 1996 to 2010. *International Communication Gazette, 76*(2), 152–176.

Schmidt, A., Ivanova, A., & Schäfer, M. S. (2013). Media attention for climate change around the world: A comparative analysis of newspaper coverage in 27 countries. *Global Environmental Change, 23*(5), 1233–1248.

Sonwalkar, P. (2008). *Global journalism.* Maidenhead: Open University Press.

Sparks, C. (2007). What's wrong with globalization? *Global Media and Communication, 3*(2), 133–155.

Statistics Norway. (2013). Statistisk sentralbyrå. Retrieved from http://www.ssb.no/en/

Thompson, J.B. (1995). *The Media and Modernity.* Cambridge: Polity Press.

UN Development Programme. (2013). *Human development reports by United Nations Development Programme.* Retrieved from https://assignmentbro.com/blog/human-development-index-and-its-components

Volkmer, I. (2015). *The global public sphere: Public communication in the age of reflective interdependence.* Cambridge: Polity Press.

Wu, H. D. (2000). Investigating the determinants of international news flow: A meta-analysis. *International Communication Gazette, 60*(6), 493–512.

CHAPTER 5

Disaster, Risk or Opportunity? A Ten-Country Comparison of Themes in Coverage of the IPCC AR5

James Painter

> Truth must be framed effectively to be seen at all. That is why an understanding of framing matters. (Lakoff 2010: p. 80)

The structure of the IPCC, which divides its analysis into three Working Groups (Physical Science Base; Impacts, Adaptation and Vulnerability; and Mitigation), testifies to the role of thematic framing in discussing the evidence about climate change. The same is true when we consider how journalism translates these findings into news coverage. Journalism itself mobilizes particular themes to make sense of the IPCC message. It also allows the framing activity of key news sources to be passed on to the public domain. By choosing the dominant thematizations and language with which to discuss IPCC findings, journalism plays a key part in constructing the dominant symbolic environment in which we discuss climate change. Such choices have consequences. As will be shown and argued below, it is a different thing to see climate change through the thematic lens of "disaster" than that of "uncertainty", "risk" or "justice".

J. Painter (✉)
Reuters Institute for the Study of Journalism, Oxford University, Oxford, UK

R. Kunelius et al. (eds.), *Media and Global Climate Knowledge*,
DOI 10.1057/978-1-137-52321-1_5

109

This chapter looks at the way AR5 coverage applied and mobilized some key thematic frames that were earlier identified as constitutive elements of public discourse. The analysis here moves from issues related to attention and access (Chaps. 3 and 4) to questions of more substantial interpretation and framing. The results about the dominant themes of reporting also further deepen the analysis of headlines (Chap. 3) and enable us to look at how some of the IPCC communication effort—such as the attempt to frame the WGII result with the language of risk (see Painter 2014, 2015b)—played out in the actual coverage. Since choosing themes is also an act of choosing a perspective, the chapter compares how different themes are emphasized in developed and developing countries. In this sense, the evidence examined here takes the question of "domestication" a step further.

FRAMES, DISCOURSES, THEMES

Research on climate change and the media tends to focus on one of three parts of the news process: the production/construction of messages, the content of media communication or the effect of such communication on audiences. Several authors have criticized the relative lack of analysis of how these three processes interact (Hansen 2011), and in particular how an examination of the whole news process (production, representation and consumption) can provide insight on how political and economic power operates (Anderson 2009; Corbett 2015). Nevertheless, the defenders of framing theory maintain that the second of these areas (in particular how a news story is written, and how some details are highlighted and others are downplayed) is of importance because it affects how individuals read and understand it. In his classic explanation of what a frame does, Robert Entman (1993) argues that a story's frame directs the reader's attention by defining a problem, stating what or who is responsible, making moral judgments and pointing to a solution.

Often inspired by Entman, scholars have chosen a wide variety of frames, both generic and specific, and applied them to media treatments of climate change (Dirikx and Gelders 2009; Nisbet 2009; Doulton and Brown 2009). It is clear from the literature that some frames have been more frequently examined than others. Variations on the "disaster/catastrophe", "uncertainty" and "opportunity" frames have been a particular focus of attention, partly because they are so often found in media treatments. For example, the disaster/catastrophe frame, which places a strong emphasis on the adverse impacts or "crisis" elements of the climate change story, such as Arctic ice melt, more drought, or more extreme weather events, was shown to be the most common frame in a comprehensive analysis of climate change coverage by the UK quality press from 1997 to 2007 (Doulton and Brown 2009). The authors found that "potential catastrophe" was by far the most common, accounting for a third of the 150 articles, while "disaster strikes" was also relatively common with around 20 articles. Mike Hulme (2009) also found that an overwhelmingly alarmist tone featured in the reporting of the WGI and WGII reports. The language of catastrophe, fear, disaster and death was an almost universal trait. Over 75 percent of items reporting on WGI and WGII fell into this category. Hulme argued this had the effect of "presenting climate change through scary, and almost pre-determined, doom-laden scenarios saturated in the language of fear and disaster, rather than as a contingent phenomenon with a malleable outcome which can be heavily influenced by policy choices" (2009: p. 126).

The "uncertainty" frame has been analyzed from several angles including the ways it can be represented as an indicator of "controversial science" where skeptics are mentioned. Probably the earliest study was carried out by Stephen Zehr (2000) who looked at the representation of scientific uncertainty around climate change in the US print media from 1986 to 1995. He concluded that uncertainty was constructed in the media by its representation in several different forms, including scientific controversy, new research topics and "expansion of the problem domain" as well as more obvious representations such as uncertainty parameters. Corbet and Durfee (2004) later argued that, in the media, science is portrayed as uncertain without ever mentioning the word "uncertain", particularly by the inclusion of competing scientific views. In her study of US newspapers in 2003–4, Antilla (2005) was primarily interested in "controversial science" where skeptics were mentioned, but she also looked at other uncertainties around the science such as ambiguous cause or effects, climate forecasts or the risk of species extinction.

The presence of skeptical arguments in the media has been extensively researched, in particular through the groundbreaking work of Maxwell Boykoff, who questioned parts of the media's tendency to give undue weight to skepticism—a process he ascribed mainly to the journalistic norm of seeking balance, which he critiques as "balance as bias" (Boykoff and Boykoff 2004; Boykoff 2007; Boykoff and Mansfield 2008). Primarily focused on the USA and the UK, his work has shown that the timing, nature and volume of skeptic presence in the media are intermittent and varied, but persistent.

The opportunity frame has been studied the least (Painter 2013), although it has been included in some journal articles. For example, it features in Doulton and Brown's study (2009) of media treatments of climate change and development, in part because climate change represented an opportunity for "social change for the poor". Nisbet (2009) includes it implicitly in his typology of eight key frames applicable to climate change, categorizing it as social progress. A cross-country and multimedia study of the 2013/4 IPCC reports incorporates opportunity as one of ten key frames, writing that "Climate change poses opportunities. Either [...] as a way to re-imagine how we live; for example, to further human development, to invest in co-benefits, [or] there will be beneficial impacts so no intervention is needed" (O'Neill et al. 2015).

More recent studies include what can be called the "explicit risk" frame where the climate challenge is presented as one of "risk management" (Painter 2013, 2014). This has been prompted by the use of such language in an increasing number of influential reports and voices in the public sphere. The "Risky Business" report launched in June 2014 is an important example of this trend (Risky Business 2014).

THE RELATIVE ABSENCE OF CROSS-COUNTRY STUDIES

Hansen (2011) was one of the first scholars to identify an urgent need to move away from single-country studies toward more comparative studies of climate change and the media. A recent meta-analysis of 133 academic articles published since 1957 found that only 10 percent of this field was longitudinal and strongly comparative (Schäfer and Schlichting 2014). This gap is beginning to be filled (Eide and Kunelius 2012; Schmidt et al. 2013), but there is a long way to go particularly when there is a pressing need to understand the media's role in explaining country differences in public opinion (Capstick et al. 2015). This chapter's original contribution is to operationalize well-established and new themes from a comparative, cross-country perspective. It is unusual to have data sets from print media in a wide range of developed and developing countries, which has allowed us to make robust comparisons of the main ways that media in each country depict the same news events, namely the release of the four IPCC WG reports.

We concentrate on print newspaper coverage of the IPCC reports in five developing countries (Bangladesh, Brazil, Chile, Indonesia and South Africa) and five developed countries (Canada, France, Japan, Norway and the UK)[1]. We were particularly interested in any differences of treatment between the two groups, although we recognize important differences within them including levels of development such as per capita income, exposure to climate change impacts, national media and political landscapes, the presence of skeptical lobby groups, the presence of prominent climate scientists, and public attitudes to climate change (Chap. 3).

The chapter focuses on the four central themes introduced above (disaster/alarm, uncertainty, opportunity and risk), and adds a fifth, climate justice, which can be seen as a variation on the blame or responsibility frame often examined in the literature (Corbett 2015; Dirikx and Gelders 2010). The "uncertainty" theme was divided into a general one evidenced by statements related to the uncertainty of scientific results and projections, and a more specific one of "skepticism" which explicitly questioned the IPCC results or the mainstream scientific consensus. Likewise, the "opportunity" theme was divided into opportunities of taking action to reduce the risks of greenhouse gas emissions (e.g. the move to low-carbon development), and opportunities from not taking action (e.g. the possibility of Arctic oil and gas exploration as sea ice melts).

The five central themes chime with the large amount of literature on how different media discourses can help or hinder public understanding, engagement or behavior change. There are different types of disaster narratives, in which the distinction between "alarming" and "alarmist"/"catastrophic" discourses is particularly important (Risbey 2008). The key point here is that doom-laden depictions of climate change are ubiquitous in the media, and yet such disaster narratives are not regarded as helpful to genuine personal engagement (O'Neill and Nicholson-Cole 2009; Hulme 2009; Rapley et al. 2014). Some scholars view emphasizing more hopeful messages, such as opportunities for low-carbon development, as a more "helpful" strategy for personal engagement than narratives of catastrophe or disaster, particularly when this is not accompanied by messages on effective actions individuals can take (Moser and Dilling 2007; Roser-Renouf et al. 2014).

[1]Although we recognize the distinction is somewhat crude, in most cases we were able to include a left-leaning/liberal/center and a right-leaning/conservative/business newspaper: Brazil *O Globo/O Estado de Sao Paulo*), Canada (*Toronto Star/Globe* and *Mail*), Chile (*La Tercera/El Mercurio*), France (*Le Monde/Le Figaro*), Japan (*Asahi Shimbun/Yomiuri Shimbun*), Norway (*Dagsavisen/Aftenposten*), South Africa (*Cape Times/Business Day*) and the UK (*Guardian/Telegraph*). In Bangladesh and Indonesia, both newspapers are considered to have no clear political leanings.

The issues around the communication of uncertainty have been amply rehearsed in other research. Several studies have shown that in certain circumstances, uncertainty can be an obstacle to public understanding, engagement or action (Shuckburgh et al. 2012; Glasgow Media Group 2012; Patt and Weber 2014). Others argue that organized skeptics, with ideological or other motivations, especially in the UK, USA and Australia, have exploited scientific uncertainties or amplified them in order to "sow the seeds of doubt" (Oreskes and Conway 2010). Consequently, the IPCC has published its own guidelines on uncertainty for its authors (Mastrandrea et al. 2010).

Evidence that policymakers and the public struggle with uncertainty has fed into an active debate as to whether, in some cases, framing climate change as one of risk (particularly for policymakers or decision-makers) is more helpful than framing it as uncertainty (Painter 2013). Many business sectors, such as the insurance and investment sectors, the military, and many politicians are used to the concept of risk. Risk language may be appropriate for some members of the general public too, as they are used to the language of betting and taking out insurance or pension policy (Painter 2015b).

The risk frame was also particularly relevant to assess in the media reporting, as the statements made by the WGII report's co-chair, Professor Christopher Field, included several references to climate change as a risk management problem. In an opening address to the IPCC in March 2014 (IPCC 2014), Field explained why thinking of climate change in terms of risk makes it easier for many to deal with:

> Climate change is really a challenge in managing risks. And it's not that we're talking about identifying particular things that are going to happen in a particular place, at a particular time. It is understanding how to be prepared in two critical ways: one is decreasing the amount of climate change that occurs, and the other is finding a way to cope as effectively as we can with the climate changes that can't be avoided.

The IPCC press release at the launch of WGII picked up on the same language (IPCC 2014). It explained two reasons why the characterization of climate change as risk management is helpful:

- It considers the full range of possible outcomes, including not only high-probability outcomes. It also considers outcomes with much lower probabilities but much, much larger consequences.
- Characterizing climate change as a challenge in managing risks opens doors to a wide range of options for solutions.

The opportunity frame is also a useful prism through which to analyze the IPCC reports as WGIII paid considerable attention to the importance of moving to renewable sources of energy, and other possible solutions to the climate change challenge.

Our research approach was heavily quantitative, one of counting the presence and salience of the five main themes and two sub-themes in each newspaper article. We use the word theme, rather than frame, as the latter would involve a wider variety of indicators (and approaches) than those captured in our coding. A theme resembles a discourse, in the sense of a line of argument or set of statements, more closely than a frame (in the fuller sense Entman described above). However, we prefer not to use the word discourse as it suggests an approach to content analysis closer to either a Foucauldian one of engaging with social-theoretical context or critical discourse analysis which usually pays close attention to the linguistic analysis of a text (Fairclough 2003).

This method was fit for purpose in that it assessed the relative presence or absence of the themes, drawing out any interesting cross-country differences. Presence was measured by a theme's appearance anywhere in an article and salience by its presence in headlines or in the opening paragraph (up to the first five lines). In addition, we were able to attain some measure of the strength of a theme's presence by coding how it appeared—as an unsourced statement, as a sourced statement, or as a direct, attributed quotation. Table 5.1 provides examples of headlines and quotes illustrating the presence of the seven themes. In some cases, multiple themes were present in the same article while in others none were present at all. The focus of the content analysis was also narrowed to articles where the four IPCC reports were the main story. This gave us a sample of 331 articles, broken down by country and report in Table 5.2.

Of the 331 articles, 135 came from the five developing countries, and 196 from the five developed countries. It is of considerable note that Brazil has the largest number of articles of any country, which reflects the large amount of media interest in climate change compared to many other countries of the world (Painter 2010, 2011). Of the developed countries, the UK had the largest sample, in part due to the significant resources *The Guardian* allocates to covering the issue (Painter 2011). In 2014, for example, *The Guardian* published more stories than any other English-language outlet, with at least 1,338 climate stories—almost four per day (Fischer 2015).

The total amount of coverage in the two newspapers selected from each of the ten countries differed sharply between WGI and WGII, and between WGIII and the SYR. In broad terms, the latter two received half

Table 5.1   Examples of themes in headlines and quotes

Theme	Headline	Quotes
Disaster	'Climate Change: just 30 years to calamity'	'By 2020, 75–250 million people will be exposed to water scarcity due to climate change'
Uncertainty	'Climate change effects unknown'	'Right now the whole debate is polarised'
Skepticism	'The world's hardly got any hotter'	'Critics say this shows carbon dioxide is not as damaging as had been claimed'
Risk	'Hundreds of millions of extra people at greater risk of food and water shortages'; 'a third of all species at risk'	'The basic economics of risk point very strongly to action'; 'society needs to take a massive insurance against this horrific prospect'
Justice	'The rich West is ruining our planet'	'The industrialised economies have created climate change, but the poorest are paying the price for it'
Opportunity from action	'Top scientists urge switch to solar power'	'There will also be geopolitical as well as commercial opportunities'
Opportunity from inaction	'Sea change for global commerce as Arctic melt transforms trade routes'	'Intergovernmental Panel on Climate Change expected to reveal that warmer sea temperatures by 2050 could open up new channels across the Arctic Northern Sea Route'

Table 5.2   Number of articles by country and report

Country	WGI	WGII	WGIII	SYR	Total
Bangladesh	2	2	2	3	9
Brazil	35	29	9	15	88
Canada	7	8	3	4	22
Chile	3	6	1	0	10
France	13	7	4	6	30
Indonesia	2	3	1	6	12
Japan	17	11	16	6	50
Norway	7	13	9	6	35
South Africa	5	5	2	4	16
United Kingdom	21	23	9	6	59
Total	112	107	56	56	331

the coverage of the former two, although there are some exceptions at the individual country level—noticeably Japan where the coverage of WGIII was unusually high.

RESULTS

Figure 5.1 shows the relative presence of the seven themes across all 331 articles in the ten countries. The disaster theme shows the highest presence at 71 percent of all articles, followed by uncertainty (52 percent), risk (38 percent), opportunity of action (20 percent), skepticism and justice (both at 15 percent) and finally opportunity of inaction at 3 percent.

Figures 5.2 and 5.3 show stronger indicators of the relative presence of the themes, as they assess salience and strength, measured by the indicators described above. In Fig. 5.2 disaster and uncertainty are again strongly present in the headlines and opening paragraphs compared to the other five themes. Disaster is particularly strong as a headline (appearing in 25 percent of all articles), compared to the next most present (uncertainty) at 9 percent. What is perhaps most interesting is that the risk theme is as present as uncertainty at 25 percent outside of the headlines and opening paragraph. Skepticism, justice and opportunity of action are all more or equally likely to be in the rest of the text than in the headlines/opening paragraph.

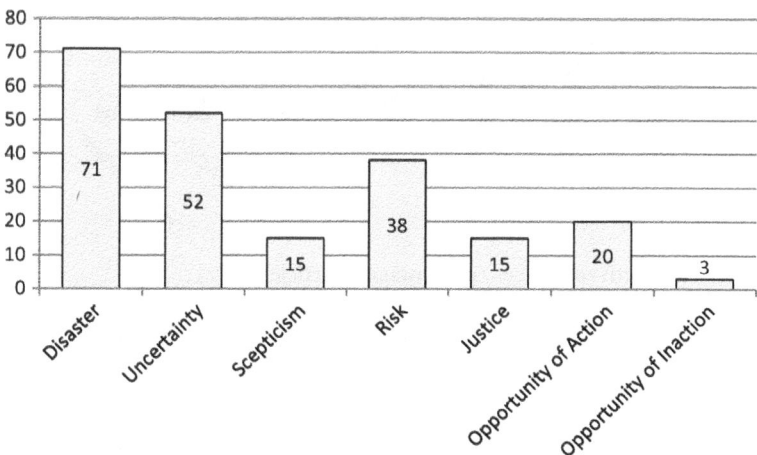

Fig. 5.1 Presence of themes (percentage of articles)

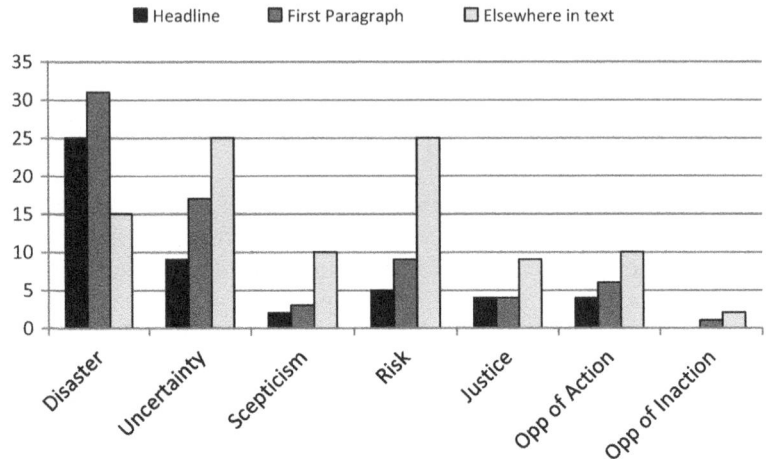

Fig. 5.2 Salience of themes (percentage of articles)

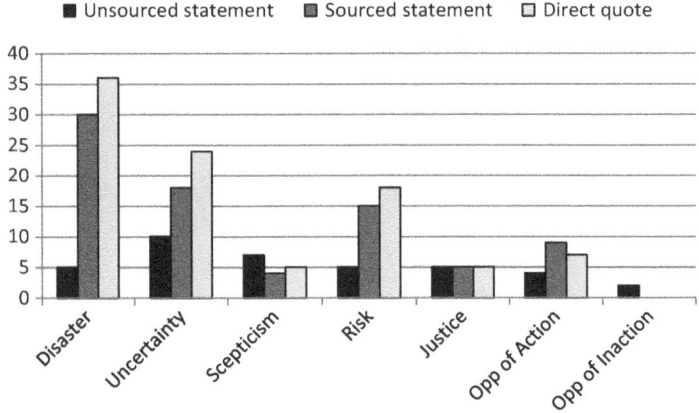

Fig. 5.3 Strength of themes (percentage of articles)

Figure 5.3 shows similar results in that disaster, uncertainty, and risk feature in the highest percentage of articles where the theme is present in a sourced statement or direct quote. In contrast, justice, skepticism and opportunity of action register in less than 10 percent of articles with a sourced statement or direct quote.

Table 5.3 divides the themes between the five developed and developing countries. It is significant that the difference in the percentage presence between the two groups is 5 percent or less for disaster, skepticism, risk and opportunity of inaction. The largest difference is found in the presence of the justice theme (15 percent), followed by uncertainty and opportunity of action (9 percent each). In the case of the justice theme, its presence is much higher in the developed countries than in the developing countries, which is perhaps counter-intuitive (Fig. 5.4).

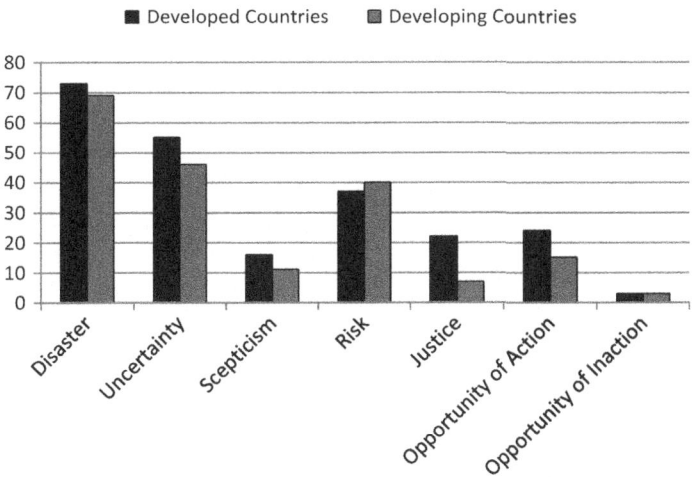

Fig. 5.4 Presence of themes: developed versus developing countries (percentage of articles)

Table 5.3 Presence of themes: developed versus developing countries (percent of articles)

Theme	Developed countries	Developing countries
Disaster	73	69
Uncertainty	55	46
Skepticism	16	11
Risk	37	40
Justice	22	7
Opportunity of action	24	15
Opportunity of inaction	3	3

THEMES FROM DISASTER TO OPPORTUNITY

The finding that the disaster and uncertainty themes are by far the most present, most salient and strongest in our samples is expected given that much of the content of the four reports focuses on the adverse impacts from a warming planet and the many uncertainties surrounding the science. It is also to be expected given that, as Table 5.2 shows, two-thirds of our sample came from coverage of the WGI and WGII reports where these themes were dominant. The skepticism theme represented only a small part of the uncertainty theme, and its presence was largely driven by the extensive space given to skeptical voices in the *Telegraphs* in the UK. The risk theme, present in 38 percent of all articles, largely reflected the widespread use of the word "risk" and, to a lesser extent, the presence of the IPCC concepts of likelihoods and certainty rankings in WGI. There was much less presence of the risk management concept pushed hard by the IPCC communication efforts around WGII.

The comparative results are more interesting. It is worth stressing that some of the coverage in the chosen newspapers in the five developing countries depended heavily on reporting by Western agencies, such as *Reuters* and *AFP*, and Western media organizations, such as the *BBC* and the *Independent*. For example, the nine articles in the Bangladeshi sample were all based on material from such organizations. Five of the sixteen South African articles and the same number of the Indonesian articles were also based on agency reports, as well as a small portion of the Brazilian sample. Hence, the division between developed and developing countries should be treated with some caution. However, it can be argued that an editor or senior journalist is still making an editorial judgment about what to include. In general, the vast majority of the articles taken from developing countries, with the exception of Bangladesh, were written by journalists based in those countries.

As mentioned above, the largest percentage difference between developing and developed countries was the climate justice theme—with far more in the latter. This theme was not present in the Bangladeshi coverage, hardly present in South Africa (1 article), Chile and Indonesia (2 articles each), and rarely present in Brazil (6 stories but only 5 percent of the sample). It is surprising that journalists in the developing world do not emphasize this aspect more, particularly as there was considerable discussion of it in the WGII and WGIII reports. In fact, the justice theme was most present in a developed country—Norway (16 articles or 46 percent), and in particular around the coverage of WGIII (7 articles). This type of frame reflects Norway's partly schizophrenic approach to the climate challenge, where its economy is dependent on oil exports but its government has invested significant amounts of money in helping developing countries' adaptation efforts. Opinion polls suggest that Norwegians consider that their country, as an oil producing nation, bears a heavier responsibility to combat climate change than others. The justice theme was also relatively highly present in Canada where six of the articles included it (27 percent of the sample). This is partly explained by one of the main environment correspondents, Raveena Aulakh, from the left-leaning *Toronto Star*, who highlighted the theme in two of her stories.

There was surprisingly little difference between developed and developing countries in the presence of skepticism (16 percent compared to 11 percent)[2]. The main driver was the two *Telegraphs* in the UK, where skeptical voices and commentary are deeply embedded (Painter and Gavin 2015). In coverage of WGI much of the skepticism was centered on discussion of the so-called climate pause, which refers to the period since the late 1990s when the global mean surface temperature has increased more slowly than during the preceding two decades. This "pause" has been used by skeptics in the USA, UK and Australia to argue that IPCC climate models are too sensitive and have exaggerated the atmospheric effects of carbon dioxide. It is noteworthy that of the 25 articles in our UK sample, where skepticism was present, 17 were published in the 2 *Telegraphs* and 7 in the *Guardian/Observer*. However, when they were present in the *Guardian*, they were refuted or balanced with a mainstream voice. The political polarization of climate change in the UK is reflected in the higher percentage of skepticism found in right-leaning newspapers than in left-leaning ones.

This is not the case in Japan. The two newspapers in our sample come from different political standpoints: *Asahi* is liberal and close to the Democratic Party of Japan, the liberal or social democratic party; *Yomiuri* is conservative and a traditional ally of the Liberal Democratic Party, the long-time ruling conservative party. Yet there was no skepticism in either newspaper despite the large sample size (50 articles). Our Japan researcher writes that "newspaper coverage hardly casts doubt on the scientific

---

[2] It should be stressed that our coding sheet did not distinguish how such voices or arguments featured. It may have been that in many articles they appeared only to be dismissed. Yet, even in this case, the presence provides a good indicator of how much presence they and/or their arguments have in a national debate.

consensus that anthropogenic emissions of greenhouse gases cause climate change. In other words, unlike the American media, the scientific controversy about the human influence on climate change has never been a salient topic in Japanese newspapers".

In our French sample, there were only two examples of skeptical voices. One of these was an article in *Le Figaro* by well-known skeptic Bjorn Lomborg. In general, there has been little inclusion of skeptical voices in the French media, except in the period between the summer of 2009 and the summer of 2010, when some skeptics were given prominence in broadcast media. Now, as our French researcher argues in reference to previous research (Comby 2015), "French journalists often follow the government agenda of wanting to make the general public more aware of the need to change behavior to tackle climate change. This is the main reason why skeptical voices are absent as they are seen as counter-productive to this aim".

The Brazilian media too historically have given short shrift to skepticism (Painter and Ashe 2012). In our sample, only seven articles included them. Our Brazilian researcher noted that skeptical arguments or voices were usually only mentioned in passing and in reference to countries where skeptics are more vocal, such as the USA.[3]

A surprising finding was the low presence of such voices in our Canadian sample—one each for the *Toronto Star* and the *Globe and Mail*. As our Canada researcher writes, "reporters at the left-leaning *Toronto Star* have tended to share the consensus that climate change requires far more action from the Harper government. By contrast, *Globe and Mail* coverage has provided more room for the oppositional voices of climate skeptics". Moreover, the Canadian sample does not include coverage from the right-leaning *National Post* where climate skepticism has been more prominent.

In Bangladesh, there is very little media space given to skepticism, which is reflected in our results (one article). However, in South Africa, *Business Day* has given significant space to skeptical voices and our sample was no exception. Two of the four articles were opinion pieces casting considerable doubt on the IPCC consensus. We did not include letters in this sample, but two letters to the same newspaper also expressed skepticism, in particular from Professor Philip Lloyd from the Energy Institute, Cape Peninsula University of Technology.

[3] For a more extensive discussion of the presence of climate skepticism in international reporting, see Painter (2011).

There was little difference between developed and developing countries in the presence of the risk theme. In most cases, the theme was the third most present after disaster and uncertainty. At 37–40 percent of the sample, this was noticeably higher than other studies have shown for the coverage of both the 2007 and 2013/4 IPCC reports (Painter 2015c). The highest percentage presence was in South Africa and Canada at 60 percent or more. In Canada, Raveena Aulakh from the *Toronto Star* consistently used the word "risk" and included direct quotations on risk from IPCC scientists. Three of the four stories in the sample written by Ivan Semeniuk, Aulakh's counterpart at the *Globe and Mail*, referred to risk, at times featuring direct quotations from IPCC scientists.

In France too, there was an above average presence of the risk theme at 50 percent. Our researcher noted prominent examples spread across the coverage in both newspapers. In Brazil, our researcher observed that "while the disaster theme was more prominent in the report released by WGI, the shift changed over the second and the third reports to a prevalent risk frame, probably due to the content of those two reports being more focused on risk management".

In general, our coding shows that the risk theme was the most prominent in the coverage of the WGII report. Of the 127 articles with some presence of this theme, 53 were in the coverage of WGII, compared to 41 for WGI, 12 for WGIII and 21 for the SYR. This may reflect the heavy usage of the word "risk" in the Summary for Policy Makers for WGII— it appeared more than 230 times in the 26-page summary. However, it was almost certainly the presence of the word "risk" which bolstered this theme (probably the weakest indicator), rather than the use of more detailed risk concepts such as "risk management". An examination of the 14 articles covering the WGII report in the South African and UK sample which included the risk theme shows that none of them contained the phrase "risk management".

Finally, the opportunity for action theme was more present in the developed countries than in the developing countries (24 percent compared to 15 percent). This was particularly true of Norway (44 percent). As our Norway researcher explains, "many of the later articles covering WGI mainly focused on the opportunities, for example how to take the lead in the 'energy revolution' towards clean energy sources. Even with WGII, where there was the highest number of catastrophe-oriented articles, all the articles stressed that we can still 'save the earth', if we act now".

It is also interesting that the opportunity for action theme was very low in Japan and Brazil, at 2 percent of articles. Our Japan researcher explains that "in the coverage of WGIII, most articles emphasized not the benefits of mitigation, but rather the challenges and costs of such actions. Particularly, they focused on the results of two degree target scenarios in AR5, drawing how difficult and tough a target it is". In the Brazilian case, this seems to be due to the fact that the newspapers in the sample usually devoted more space to criticism of the government's inaction in that area, rather than showcasing the potential benefits of reacting to climate change.

CONCLUSIONS

We should be wary of drawing too strong a set of conclusions from our study. First, as we have discussed, the sample from the five developing countries was dominated by Brazil (which represented 65 percent of the sample). Secondly, sampling from a wider selection of media in each country might have produced different results. For example, in South Africa, the non-English print press may have provided different conclusions. Television, which in many developing countries is much more widely used than print media, may have given different results. Thirdly, we restricted the number of themes to five major and two minor ones. Other studies of the IPCC reports have included a greater variety of themes to include such frames as "settled science", "political struggle", "role of science", "security", "morality and ethics", and "health" (O'Neill et al. 2015). Finally, the construction and delineation of the seven themes was problematic. The "settled science" frame is particularly important as much of the narrative around WGI was that scientists were more certain than ever about the causes of climate change. Indeed, although we coded this as "uncertainty", a strong argument can be made that the message a reader received would have been more along the lines of "scientists are very sure that...", rather than "scientists still cannot be sure that...". The other strong theme to emerge from our sample was the "need or call for action" which was prominent in many articles but not coded separately.

However, we were particularly interested in comparisons between media treatments in developed and developing countries. The selection of our themes was consistently applied across all ten countries, which makes our results sufficiently robust. The predominance of the disaster theme in all of the countries is closely linked to the wider coverage given in our sample to the WGI and WGII reports. Some studies have commented on the relative lack of coverage of the WGIII report which also helps explain the relative absence of the opportunity of action theme (Painter 2014; O'Neill et al. 2015). This is prima facie surprising because the WGIII report provides several interesting journalistic angles. As O'Neill has written, "WGIII contains much newsworthy material (e.g. What energy future do we want? Should energy provision be more equitable? Should consumption of red meat be restricted?) ... All of these seem compelling issues for news media, certainly in comparison to the physical science focus of WGI" (O'Neill et al. 2015). The explanation probably lies in a combination of the way WGIII was released, soon after WGII, (possibly making it vulnerable to "climate fatigue" among journalists), the date of the release (a Sunday—April 13th), the IPCC's presentation of the report, and the relative lack of compelling narratives and human interest stories to illustrate it.

The relatively lower presence of the other themes, and in particular, opportunity of action and risk, is broadly consistent with other studies of media coverage of the AR5 reports (Painter 2014, 2015a; O'Neill et al. 2015). Our cross-country study reveals little variation with studies that have primarily focused on coverage in the developed world. This may well be a product of the nature of the story as a news event, where the "push factors" framing the story—messages and headlines emanating from the IPCC communication effort, the reliance on Western agency reporting, the relative absence of different story lines for each report—strongly shapes the dominant themes in a story whatever the national context in which the media are covering it. However, this does mean that often the messages about climate change a reader receives are those of fear and uncertainty, rather than hope, opportunity, justice or the possibility of effective risk management. It will remain a massive challenge for the IPCC and other organizations to shift the dominant narratives in the future.

BIBLIOGRAPHY

Anderson, A. (2009). Media, politics and climate change: Towards a new research agenda. *Sociology Compass, 3*, 166–182.

Antilla, L. (2005). Climate of scepticism: US newspaper coverage of the science of climate change. *Global Environmental Change, 15*, 338–352.

Boykoff, M. T. (2007). Flogging a dead norm? Newspaper coverage of anthropogenic climate change in the United States and United Kingdom from 2003 to 2006. *Area, 39*(2), 470–481.

Boykoff, M. T., & Boykoff, J. M. (2004). Balance as bias: Global warming and the US prestige press. *Global Environmental Change, 14*(2), 125–136.

Boykoff, M. T., & Mansfield, M. (2008). "Ye olde hot Aire": Reporting on human contributions to climate change in the UK tabloid press. *Environmental Research Letters, 3*(2), 1–8.

Capstick, S., Whitmarsh, L., Poortinga, W., Pidgeon, N., & Upham, P. (2015). International trends in public perceptions of climate change over the past quarter century. *WIREs Climate Change, 6*, 35–61.

Comby, J. (2015). *La question climatique: sociologie d'un processus de dépolitisation*. Paris: Raisons d'Agir, coll. Cours & Travaux.

Corbett, J. (2015). Media power and climate change. *Nature Climate Change, 5*, 288–290.

Corbett, J., & Durfee, J. (2004). Testing public (un)certainty of science: Media representations of global warming. *Science Communication, 26*, 129–151.

Dirikx, A., & Gelders, D. (2009). Global warming through the same lens: An explorative framing study in Dutch and French newspapers. In T. Boyce & J. Lewis (Eds.), *Climate change and the media* (pp. 200–210). New York, NY: Peter Lang.

Dirikx, A., & Gelders, D. (2010). Ideologies overruled? An explorative study of the link between ideology and climate change reporting in Dutch and French newspapers. *Environmental Communication, 4*(2), 190–205.

Doulton, H., & Brown, K. (2009). Ten years to prevent catastrophe? Discourses of climate change and international development in the UK press. *Global Environmental Change, 19*, 191–202.

Eide, E., & Kunelius, R. (2012). *Media meets climate: The global challenge for journalism*. Gothenburg: Nordicom.

Entman, R. M. (1993). Framing: Toward clarification of a fractured paradigm. *Journal of Communication, 43*(4), 51–58.

Fairclough, N. (2003). *Analyzing discourse: Textual analysis for social research*. London and New York: Routledge.

Fischer, D. (2015). Back in the headlines: Climate coverage returns to its 2009 peak. *The Daily Climate*. Retrieved from http://www.dailyclimate.org/tdc-newsroom/2015/01/climate-change-coverage-2014

Glasgow University Media Group. (2012). *Climate change and energy security: Assessing the impact of information and its delivery on attitudes and behaviour*. London: UK Energy Research Council.

Hansen, A. (2011). Communication, media and environment: Towards reconnecting research on the production, content and social implications of environmental communication. *International Communication Gazette, 73*(1–2), 7–25.

Hulme, M. (2009). Mediated messages about climate change: Reporting the IPCC Fourth Assessment in the UK print media. In T. Boyce & J. Lewis (Eds.), *Climate Change and the Media* (pp. 117–128). London: Peter Lang.

IPCC. (2014, March 31). *IPCC report: A changing climate creates pervasive risks but opportunities exist for effective responses*. Geneva, Switzerland: IPCC. Retrieved from http://www.ipcc.ch/pdf/ar5/pr_wg2/140330_pr_wgII_spm_en.pdf

Lakoff, G. (2010). Why it matters how we frame the environment. *Environmental Communication, 4*(1), 70–81.

Mastrandrea, M. et al. (2010). *Guidance note for lead authors of the IPCC Fifth Assessment Report on consistent treatment of uncertainties.* Released by IPCC. Retrieved from https://www.ipcc.ch/pdf/supporting-material/uncertainty-guidance-note.pdf

Moser, S. C., & Dilling, L. (Eds.). (2007). *Creating a climate for change: Communicating climate change and facilitating social change.* Cambridge: Cambridge University Press.

Nisbet, M. (2009). Communicating climate change: Why frames matter for public engagement. *Environment Magazine, 51,* 12–23.

O'Neill, S. J., Kurz, T., Williams, H. T., Wiersma, B., & Boykoff, M. (2015). Dominant frames in legacy and social media coverage of the IPCC Fifth Assessment Report. *Nature Climate Change, 5,* 380–385.

O'Neill, S. J., & Nicholson-Cole, S. (2009). Fear won't do it: Promoting positive engagement with climate change through imagery and icons. *Science Communication, 30,* 355–379.

Oreskes, N., & Conway, E. (2010). *Merchants of doubt: How a handful of scientists obscured the truth on issues from tobacco smoke to global warming.* New York, NY: Bloomsbury Press.

Painter, J. (2010). *Summoned by science: Reporting climate change at Copenhagen and beyond.* Oxford, UK: Reuters Institute for the Study of Journalism.

Painter, J. (2011). *Poles apart: The international reporting of climate scepticism.* Oxford, UK: Reuters Institute for the Study of Journalism.

Painter, J. (2013). *Climate change in the media: Reporting risk and uncertainty.* Oxford, UK: I.B. Tauris and Reuters Institute for the Study of Journalism.

Painter, J. (2014). *Disaster averted? Television coverage of the 2013/14 IPCC's climate change reports.* Oxford, UK: Reuters Institute for the Study of Journalism.

Painter, J. (2015a). Disaster, uncertainty, opportunity, or risk? Key messages from the television coverage of the IPCC's 2013/14 reports. *METODE Science Studies Journal, 85,* 73–79.

Painter, J. (2015b). The effectiveness of the IPCC communication: A survey of (mainly) UK-based users. *Advance paper submitted to the IPCC Expert Meeting on Communication.* IPCC, Oslo, Norway. Retrieved from http://www.ipcc.ch/meeting_documentation/meeting_documentation_ipcc_workshops_and_expert_meetings.shtml

Painter, J. (2015c). Taking a bet on risk. *Nature Climate Change,* , 286–288.

Painter, J., & Ashe, T. (2012). Cross-national comparison of the presence of climate scepticism in the print media in six countries, 2007–2010. *Environmental Research Letters, 7*(4), 1–8.

Painter, J., & Gavin, N. (2015, January 27). Climate skepticism in British newspapers, 2007–2011. *Environmental Communication,* 1–21. doi:10.1080/17524 032.2014.995193

Patt, A. & Weber, E. (2014). Perceptions and communication strategies for the many uncertainties relevant for climate policy. *WIREs Climate Change, 5,* 219–232.

Rapley, C., de Meyer, K., Carney, J., Howarth, C., Clarke, R., Smith, N., et al. (2014). *Time for a change? Climate science reconsidered.* Report of the UCL Commission on Communicating Climate Science, University College, London.

Risbey, J. (2008). The new climate discourse: Alarmist or alarming? *Global Environmental Change, 18,* 26–37.

The Risky Business Project. (2014). *Risky business: The economic risks of climate change in the United States.* Retrieved from http://riskybusiness.org/uploads/files/RiskyBusiness_Report_WEB_09_08_14.pdf

Roser-Renouf, C., Maibach, E. W., Leiserowitz, A., & Zhao, X. (2014). The genesis of climate change activism: From key beliefs to political action. *Climatic Change, 125,* 163–178.

Schäfer, M., & Schlichting, I. (2014). Media representations of climate change: A meta-analysis of the research field. *Environmental Communication, 8*(2), 142–160.

Schmidt, A., Ivanova, A., & Schäfer, M. S. (2013). Media attention for climate change around the world: A comparative analysis of newspaper coverage in 27 countries. *Global Environmental Change, 23*(5), 1233–1248.

Shuckburgh, E., Robison, R. & Pidgeon, N. (2012). Climate science, the public and the news media: Summary findings of a survey and focus groups conducted in the UK in March 2011. Released by Living with Environmental Change web page. Retrieved from http://nora.nerc.ac.uk/500544/

Zehr, S. (2000). Public representations of scientific uncertainty about global change. *Public Understanding of Science, 9,* 85–103.

CHAPTER 6

Journalism, Climate Change, Justice and Solidarity: Editorializing the IPCC AR5

Anna Roosvall

Justice is a significant undercurrent in journalism and in international politics on climate change. The questions "who is to blame?" and "who is responsible?" are recurrently evoked in both areas. Yet the issue of justice is seldom explicitly discussed or problematized in these contexts. Today theories of justice are being reconstructed in order to relate them to a globalizing world (Fraser 2008, 2014; Nash 2014; Owen 2014; Sen 2009). The old conceptualizations were generally aligned with the nation-state (Rawls 1971); however, increasingly global connections, transnational exchanges and global risks, such as climate change, call for a reformulation. How does journalism on the international politics of climate change relate to these calls? The international politics of climate change are most pertinently discussed and *evaluated* in editorial sections and equivalent recurring special opinion sections dedicated to the environment, which constitute the research material in this theoretical and explorative study.

Our contemporary world is marked by competition over *economic*, *natural* and *information* resources. For justice to be implemented when competition is fierce, responsibility and solidarity are necessary. This chapter therefore theorizes not only justice, but also the role responsibility plays for justice and—in a specific section—solidarity. It is largely theoretical,

A. Roosvall (✉)
Department of Media Studies, Stockholm University, Stockholm, Sweden

© The Author(s) 2017   129
R. Kunelius et al. (eds.), *Media and Global Climate Knowledge*,
DOI 10.1057/978-1-137-52321-1_6

discussing the need for a retheorizing of justice in a globalizing world, relating this to media studies, and developing an analytical approach specifically suited for opinionated material. Subsequently it explores notions of justice and attitudes toward it in a small sample of articles from editorial and special opinion sections following the releases of the IPCC AR5 in 2013–2014. The chapter reveals how climate change is understood as a political issue from a justice perspective in a globalizing world. This is accomplished by analyzing pieces from the aforementioned sections, particularly in high-income countries[1] with privileged access to economic and information resources who have historically been the greatest polluters. The research questions are, how does editorial and specialized opinionated output in high-income countries approach issues of (in)justice in relation to climate change? How do they relate geographical scales (global, national, local) and temporal scales (past, present, future) to climate (in)justice? Do the articles prescribe how to remedy injustice? Do they relate solidarity to climate justice? Who are termed responsible for (in)justice in the discourses of the articles?

THEORIZING JUSTICE AND SOLIDARITY IN A GLOBALIZING WORLD

Justice

In his classic definition of justice Rawls (1971) calls it "the first virtue of social institutions" (p. 3). He underlines that justice should not only be defined as social but also connected to institutions which are, in his terms, national. Later theorizing has questioned both these aspects. First, the social has gradually appeared as *one* important aspect of justice rather than *the* most important, adding for instance environmental and climate justice (Roosvall and Tegelberg 2015). Struggles for climate justice turn climate change into an issue concerning ethics, global equality, human rights and historical responsibility (de Onís 2012; Page 2006; Shepard and Corbin-Mark 2009; Shue 2014). Second, geographic claims must be reconsidered (Fraser 2008, 2014; Nash 2014; Owen 2014; Sen 2009) since justice theories were formed in relation to institutions governed by nation-states (Rawls 1971), before the intensification of globalization from the 1980s onwards, and prior to when the transnational phenomenon of climate change was widely

[1] For information on high-income countries, see http://data.worldbank.org/income-level/OEC

recognized. Young (2013: p. 123), for instance, argues for a "social connection model" of justice that expands responsibility across national borders.

Fraser's (2008) most basic definition of justice is "parity of participation" (p. 16). A pertinent way to illustrate this is to elaborate on understandings of injustice, which have varied over the years. After WWII, remedy in the form of *redistribution* dominated such understandings until gradually replaced by *recognition* in the 1990s (Fraser 2000). This was followed by a struggle for the restoration of redistributive action, particularly by the left. Fraser argues that redistribution, which remedies economic damage, and recognition, which remedies cultural insult, are both needed (Fraser 2000, 2008; Fraser and Olson 2008). Processes of globalization and debates with other scholars concerning redistribution versus recognition (Fraser and Honneth 2003; Fraser and Olson 2008; Nash 2014), subsequently led Fraser to introduce a third, complementary approach: *representation*, that is, *political representation*. She, in turn, adds to these three moral remedies (redistribution, recognition and representation) a geographical dimension distinguishing between *affirmative* and *transformative politics of framing* to counteract the worst injustice of all in our globalizing world: *misframing*. Misframing occurs when an inappropriate scale is imposed upon a case of injustice, such as when a national scale and solution are employed to address a transnational source of injustice. Young (2013: p. 137) makes a similar argument concerning collective responsibility for justice, claiming that the national scale cannot be used unless actions are clearly connected to it. This is highly relevant to the politics of climate change where causes of injustice stretch beyond national scales; exemplified, for instance, by climate change effects that are spread across several nations in the Polar Regions or in the Amazon rainforest. An affirmative politics of framing looks beyond national scales in certain cases, but does not question national framing as such. A transformative politics of framing looks further, questioning the use of the national scale in remedies to transnational injustices in *any* case. The global justice movement and indigenous groups engaged in environmental activism have worked with the latter approach, questioning the very structure of international politics by trying to disconnect the geographical scale of the nation from the moral scale of justice.

Mirroring Fraser's calls for a retheorizing of justice in a globalizing world, Couldry discusses justice in relation to digitalization as well as globalization, identifying the need for a new *media* ethics (see also Deuze 2005; Rao and Wasseman 2015). Couldry (2012: p. 181) discusses environmental ethics in relation to global environmental damage to demonstrate the relevance of what I, with Fraser (2008), call a global reframing. Media, as the main framing agents in contemporary societies, hold great potential to contribute to or even lead such a reframing effort. The media also carry the potential to make injustices visible. In this regard, Silverstone (2007) emphasizes that media as institutions can and do act justly as well as unjustly, for instance, by "denying or distorting the voices of others" (p. 144). In a globalizing world, the severity of such foreclosing deepens when there is an increase in audiences who may be theoretically able to hear but who cannot practically do this due to the systematic exclusion of certain voices and perspectives (ibid.).

Climate justice, encompassing equality, human rights and historical responsibility, can be delimited as intranational, international, transnational or global. *Intranational* climate injustice is comparable to environmental racism; emphasizing that pollution and waste are often directed toward poor, non-white communities (Bullard 1990). *International* climate justice concerns "what kinds of allocations of costs between rich nations and poor nations it would be acceptable for a poor nation to accept" when climate change is tackled (Shue 2014: p. 27). This relates to one of the main challenges at UN climate summits, where in 2013 at the Warsaw summit, for example, the G77 nations and China walked out of negotiations to protest the lack of compensation for so-called developing countries facing extreme climate events (a measure opposed by the EU, Australia, the USA and other high-income countries). *Transnational* climate justice focuses on the effects of climate change on transnational groups, such as certain indigenous peoples (de Chavez and Tauli-Corpuz 2009; Roosvall and Tegelberg 2015). Finally, *global* climate justice combines the above mentioned spatially limited forms into a view that spans across the planet.

There is a dearth of research on journalism and climate justice.[2] Roosvall and Tegelberg (2012, 2015) deal with intranational and transnational climate justice and indigenous peoples. Rhaman (2014) considers international climate justice by analyzing how Bangladeshi journalism approaches climate justice conceptually in coverage of UN climate summits. He concludes that while news articles only relate to climate justice "connotatively", opinionated articles relate to it "denotatively", that is explicitly. These findings motivate the focus here on global climate justice in opinionated articles on the IPCC reports, as well

---

[2] There has been more research on social media and climate justice (Askanius and Uldam 2012; Uldam and Askanius 2013) and on climate change protests (Segerberg and Bennett 2011; Ytterstad and Russell 2012).

as the role international, intranational and transnational climate justice play in constituting global scales.

Responsibility for justice also expands in time. Young (2013) claims responsibility for justice should mainly be discussed in the *future* temporal scale (see also Page 2006), whereas Nussbaum (2013) underlines the importance of past and historical justice. In climate change discussions, the future tense is frequently evoked to speak of the responsibility present generations hold toward future generations. Responsibility for the past comes up, for example, in "the polluter pays" principle (Caney 2005), which deems those with the highest levels of historical emissions most responsible. However, when it is not combined with historic responsibility, the same principle can be used to support claims that nations, like China, with the highest emissions today, are the most responsible for making changes. This represents another crucial controversy in international climate change negotiations: how far responsibility extends, where it ends and whether previously poor countries should be allowed to "catch up" with major emitters of the past.

Parallel to the notion of responsibility is the question of who is affected by injustice. A distinction between the *all-affected* and the *all-subjected* principle helps to elaborate this (Fraser 2008; Owen 2014). The all-affected principle is used to discuss justice by defining who is affected by an unfair situation or action, such as flooding, ice melting, droughts and so on. The all-subjected principle instead encompasses all who are implicated in the structures to which the injustice is connected. As Fraser (2008) puts it, "what turns a collection of people into fellow members of a public is not their shared citizenship, or co-imbrication in a casual matrix, but rather their joint subjection to a structure of governance that set the ground rules for their interaction" (p. 96). When it comes to a global issue like climate change many more people thus become part of the justice issue; everyone can in a sense be considered subjected.

Young (2013) elaborates on what can be expected of us in relation to issues of injustice: "we can and should be criticized for not taking action, not taking enough action, taking ineffective action, or taking action that is counterproductive" (p. 144). Those more responsible for taking issue with injustice are, according to Young (2013: p. 145), persons privileged by the system as well as victims of it, since they are directly affected. Ability and power are also important factors in determining who are most responsible for action (Young 2013: p. 147). This is where the news media become important. The power to take issue depends largely on access to media, which function as channels for discursive action, spreading the message of injustice and outlining the remedies required. Today's media ecologies are however harsh ones for "victims" such as indigenous activists aiming to spread global awareness of climate change effects in their communities (Roosvall and Tegelberg 2015). These groups feel the mainstream media must represent them in order for them to be heard widely, while the mainstream media largely ignore indigenous voices and perspectives; particularly when it comes to redistribution and political representation. Recognition of status is rare as well. Thus, it seems that placing so much responsibility on victims, as Young does, is unfair in the current media landscape. Determining how attitudes toward justice appear in the centers of mainstream media power, in turn, forms part of what motivates this study of editorializing in journalism from high-income countries.

While responsibility for justice seems difficult and overwhelming this is not ground for ignoring it. On the contrary, such an approach would be conservative in an unjust world (Young 2013: p. 123). Thus, we cannot dwell on the overwhelming feeling; the remedy here is to move to action (Young 2013: p. 124). A responsible person "tries to deliberate about options before acting, makes choices that seem to be the best for all affected, and worries about how the consequences of his or her actions may adversely affect others" (Young 2013: p. 25). Assigning responsibility to someone is to say they have a job to do; they have to act (Nussbaum 2013; Young 2013; Roosvall 2014a). *Political* responsibility consists, for instance, in monitoring social institutions and their effects "to make sure they are not grossly harmful, and maintaining organized public space where such watching and monitoring can occur and citizens can speak publicly and support one another in their efforts to prevent suffering" (Nussbaum 2013: p. xv). One of the media's main roles is such monitoring. Silverstone (2007: p. 11) points out that technologies, like the media, disconnect as well as connect. That is, they often undermine the expectation of responsibility that action in face-to-face settings would demand. Hence, focusing on how media represents the other is a key issue for any project seeking a more virtuous ethical public space.

Solidarity

Questions of responsibility connect the justice debate to the notion of solidarity. Responsibility for justice is central in justice theorizing while solidarity is sometimes associated with collective responsibility (Smiley 2010). Although journalism and justice have rarely been studied together, the intersection of the two is often implied in research on journalism and solidarity (Chouliaraki 2011, 2013; Roosvall 2014a). Eide (2012) relates *editorial* solidarity to climate change in a study of the famous *Guardian* editorial published by 56 newspapers around the world at the opening of the Copenhagen climate summit in 2009. In this chapter, solidarity is not only understood to signify collective responsibility, but to further encompass action, reciprocity, empathy and restitution (Roosvall 2014a). In the following, I briefly elaborate on these aspects.

Solidarity is connected to a political worldview (Fenton 2008) and materialized through words and *actions* of collective responsibility. As Liedman (1999) summarizes, solidarity is not merely a "brotherhood of phrases" but a "brotherhood of action". Actions can however be discursive if we "recognize the *performative* power of humanitarianism—its power, that is, to constitute vulnerable others as worthy or unworthy of our emotions" (Chouliaraki 2013: p. 23, emphasis added). These *performances* of solidarity frequently take place in the media.

Solidarity as (discursive) action is illustrated further by Chouliaraki's (2011) understanding of solidarity as agonism: "the communication of human vulnerability as a political question of injustice that can become the object of our collective reflection, *empathic* emotion and transformative *action*" (p. 377, emphasis added). Chouliaraki evokes yet another term used to define solidarity: empathy. Kössler and Melber (2007: p. 32) include empathy in a list of aspects that constitute solidarity: identity, substitution, complementarity, reciprocity (including empathy), affinity and restitution. Reciprocity and restitution are part of the focus here since they are closely related to justice and journalism.

Reciprocity signifies a dialogic *mode of communication* that addresses exchanges of experiences/assistance in its communicative action (Roosvall 2014a). Thus, solidarity is not a one-way form of communication, like charity, but a two-way mode. *Collective* action is expected. Solidarity as a mode of communication however stretches beyond dialogue because the exchange is supposed to be similar or equal, which is not true of all dialogue (Roosvall 2014a), and because it includes actions specifically directed toward injustice. Thereby, it surpasses ideas of dialogue leading to peace and agreement since the mending of injustice will not be agreed upon by everyone. This is inherent in the agonistic approach.

Restitution, in turn, connects to the *representational aspect of communicative action*, rather than to modes of communication. It refers to the recognition of historical injustice as a basis for action (which includes the restitution of more equal power relations). It is retroactive and redistributional and thereby related to advocacy of *historical* responsibility in justice theory (Nussbaum 2013).

METHODOLOGICAL APPROACH AND RESEARCH MATERIAL

The research sample consists of articles coded as editorial material in the *MediaClimate* sample of newspapers (Chap. 1). Editorial material, as it appears in various countries, is well suited for studies of climate politics and justice in the media: it is the journalistic genre closest to the field of power. According to Kunelius, it is editorial content that maps the diversity of legitimate opinions (2012: p. 37). The imagined audience is national elites (Eide 2012). Kunelius adds that editorials "carve out a conceptual space and argue within it, thus describing the key possibilities and potentials that the current debate on climate politics offer...[they] ...speak with and comment on the *dominant language* of politics" (2012: p. 37). Nord (1998, 2001) describes the editorial as simultaneously a self-produced official statement of the newspaper and the "soul" of the newspaper. It is an evaluative text that expresses the newspapers' political views; it criticizes and favors. Editorial content is generally skewed toward opinions that a newspaper readership is assumed to agree with. Thereby, material from the editorial sections produced around the releases of the IPCC reports provides excellent material for exploring public discourse on climate change. Articles termed editorials do not appear in the exact same way in all the countries included in this sample. Generally, editorials are un-signed, but some countries will also term signed articles editorials when they are published on a specific page dedicated to opinionated journalism. In the US material, for instance, a special online opinion section called *Dot Earth* was coded as editorial material. This regular opinion page is exclusively dedicated to the environment, notably not for featuring the invited opinions of diverse writers, but instead signified by one recurrent voice (Andrew Revkin). Despite the variety in this sample of editorial material, a point of common ground is taken as the starting point for this analysis: each of these articles *editorializes* climate politics in relation to the release of the AR5 report. The specificities of each analyzed article (whether it is signed or not etc.) are explained in footnotes associated with each of the closely analyzed articles.

Newspapers in the USA, UK, Canada, Brazil, Chile, Norway, Finland, Sweden, France, Germany, Bangladesh, Japan and Australia featured such material (Fig. 6.1). Thus, the majority of this type of material was published in high-income countries (apart from Brazil and Bangladesh). All these articles are first categorized according to themes in the headlines. Then, a selection of articles from five of the high-income countries are examined by combining a critical linguistic approach with an operationalization of the previously discussed concepts from Fraser's (2000,

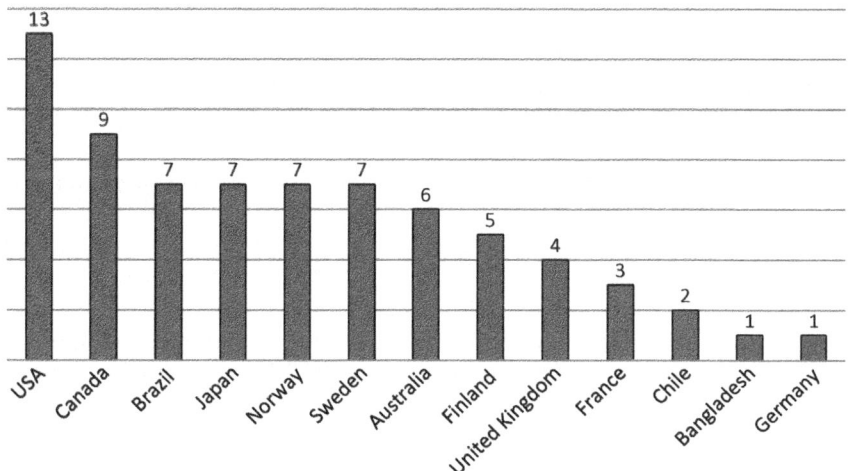

Fig. 6.1 Number of editorials per country (72 in total)
Note: The sample includes three Swedish newspapers since the Swedish material was taken as a starting point for this study. *Aftonbladet,* Sweden's largest daily, was selected as an additional newspaper since it was recently included in an associated study (Larsson, 2015)

2008, 2014) and Young's (2013) theorizing on (in)justice. As previously mentioned, the primary aim of this chapter is theoretical and exploratory. Hence, the items examined here were chosen because they provide opportunities to explore how editorializing in journalism on the IPCC AR5 relates to justice, responsibility and solidarity; especially in view of the retheorizing being done in these areas in the context of globalization. Therefore, the articles selected for closer analysis (one per selected country, see further on) were chosen because their headlines indicate through word choices that they will in some way relate to issues of justice, responsibility and solidarity, possibly on a global scale. Due to its limited scope (mainly based on one article from each country), the analysis should not be taken as the basis for broader, empirically grounded conclusions regarding the selected countries.

Exploring Spatial and Temporal Conceptions of Justice

In order to answer the research questions, Fraser's theoretical concepts are operationalized by identifying (see also Roosvall 2014b; Roosvall and Tegelberg 2015):

Remedies for injustice

- whether the articles call for/relate to redistribution
- whether recognition is evoked and, if so, how (is identity or status acknowledged?)
- whether the articles advocate and/or discuss political representation in relation to justice

Spatial aspects

- statements relating to different scales; local, national and global, concerning climate change (effects), as well as international climate change politics

Temporal aspects

- statements relating to different times; the past, the present and the future, concerning climate change (effects), as well as international climate change politics

Who the injustice concerns

- all-affected?
- all-subjected?

Exploring Attitudes (Modality) Toward Justice

The critical linguistic approach builds on Fowler's (1991) modality analysis. Modality derives from grammar and is often expressed through modal verbs that state attitudes toward something. Fowler (1991) lists four forms of modality: desire, obligation, permission and truth/probability. This study examines attitudes toward justice as expressed through these four modality forms. It adds the modality of solidarity, which emphasizes that there are particular solidaritarian modes of communication including calls to action, a will to provide restitution for past wrongs (inequalities) as well as empathy and reciprocity.

The analysis of these five modalities is concentrated on parts of the articles that contain statements concerning climate change and (in)justice. The modalities are used here to identify the following attitudes toward justice:

- obligation, the attitude that something *ought to* be done;
- desire, the attitude that it would be good if something was done, but not absolutely necessary;
- permission, an attitude acknowledging something, that might be viewed as questionable, *could* be done;
- attitudes toward truth and probability
- attitudes that express solidarity

Desire and *obligation* convey the strength of the arguments concerning climate injustice. Are remedies "only" desired or are they absolutely necessary (obligation)? *Permission* could appear, for instance, as permission to emit certain amounts of greenhouse gases. The *truth and probability* attitude is always at play in journalism. The latter specifically connects to the probability measurements in the IPCC reports, while also signaling the degree of urgency. Higher probability of acute climate change effects implies a higher degree of urgency, which may lean more toward obligation than desire. Concurrently, if the truth modality is low the obligation modality may be demoted. In line with critical linguistic approaches not only what is present, but also what is absent is considered in the analysis.[3]

ATTITUDES TOWARD JUSTICE IN THE SELECTED ARTICLES

The headlines from each of the articles coded as editorials were collected in order to identify the main themes. In relation to the understanding of climate justice (1) as concerning ethics, global equality, human rights and historical responsibility, and (2) as issues of "what kinds of allocations of costs between rich nations and poor nations it would be acceptable for a poor nation to accept" (Shue 2014: p. 27), the identification of themes was specifically attentive to conflicts, responsibility and restitution as well as to assessments of these issues. Eight themes were identified: (1) conflict, struggle or polarization; (2) overcoming conflict, agreement, reconciliation, "break-through"; (3) assessing domestic policy; (4) planet/globe, nature; (5) meta-debate; (6) IPCC report emphasis; (7) effects; and (8) other.

The first two themes represent two sides of the same coin. Together with "assessing domestic policy" these themes cover most of the headlines. In these headlines, groups or nations are pitted against each other, an antagonism was seen as being resolved, or a domestic policy was compared to policies in other nations. Articles conveying these themes have been selected for further analysis. The analysis aims to interrogate discussions

---

[3] Images are not considered here since the analysis focuses on linguistics

about climate politics spurred by the release of the AR5 report. We know from analysis of themes in all genres that justice themes are more pronounced in coverage of "developed countries" (Chap. 4). Although this may seem counter-intuitive at first, it underlines the need to understand more qualitatively *how* justice arguments are articulated in nations that are historically responsible for the most emissions.

The exploratory study of high-income countries begins with Sweden, selected because it is a country that pays high attention to climate change effects as a threat to humankind (Global Challenges Foundation 2014). It then moves on to articles from G7 countries, namely the USA (selected because it is often viewed as a global laggard in climate politics), Japan (selected because it is a country with high awareness of environmental issues, especially after Fukushima), Canada (selected because it is viewed internationally and domestically as a nation engaging in unfriendly climate politics) and France (selected because it is the host of the 2015 climate summit). One article from each of these countries is analyzed, placing emphasis on (overcoming) conflict, assessments of domestic policy and/or calls for (global) action in the headline. Other editorial items from the same countries are occasionally referred to in the analysis as secondary articles.

"The Climate Threat: A Political Problem". Aftonbladet, *Sweden*[4]

This editorial chronicle includes a subheading with the phrase "everyone must sacrifice". It implies a call for justice in that everyone must carry the burden. This evokes the all-subjected, rather than the all-affected, principle with regard to who is implicated in this injustice. In a solidaritarian way, we should all sacrifice something, regardless of whether we are responsible for direct causes or not. The article does not however argue for affirmative action or restitution. It claims climate change is a "political problem", with "hindrances [that] are political. Measures have to be taken now", but it does not discuss justice in terms of political representation, redistribution or recognition. Responsibility for our children and grandchildren is mentioned—"we cannot tell our children and grandchildren that we did not know"—thus the future temporal aspect of justice is implied. As is the spatial aspect:

[4]Persson, I. (2014). The climate threat—A political problem. *Aftonbladet*, November 3. Retrieved from http://www.aftonbladet.se/ledare/ledarkronika/ingvarpersson/article19794981.abhttp://www.aftonbladet.se/ledare/ledarkronika/ingvarpersson/article19794981.ab. This piece is presented as an editorial chronicle, produced by an editorial writer and published under the heading "Editorial".

This is furthermore about trust. The world's leaders know that emissions must decrease, but know at the same time that their own country's sacrifices are pointless if other countries do not do their part as well. In a world as unjust as ours, it is indeed not easy to assess how big that part should be.

The editorial chronicle addresses justice explicitly, arguing that the issue must be approached globally. It goes beyond the old nation-state boundaries by aiming for broader, more inclusive geographic scales of justice. At the same time a "you first" approach is evoked; that is countries are presumed to be waiting to see what other countries will do. Hence, while the all-subjected principle is implied by "everyone must sacrifice", it is evident that conflicts over fairness regarding "how big that part should be" can potentially halt the collective action of solidarity.

The "everyone must sacrifice" call moreover includes a strong obligation modality, emphasizing that something "must" be done. "We cannot claim we did not know" concurrently expresses a permission modality: we do not have permission to claim we did not know. This strong language, however, is not combined with any specification of particular remedies for injustice. Hindrances are political, but political remedies are not mentioned. We learn that "politicians and diplomats should do their job" (again, a strong obligation modality), but if the ground for political negotiations is unjust, as indicated in this editorial, then specified political remedies for injustice (i.e. representation) are needed.

"In Joint Steps on Emissions, China and US Set Aside 'You First' Approach on Global Warming". New York Times, USA[5]

This editorial piece is written by Andrew Revkin, a key figure in climate change journalism and commentary in the USA. It appears on *Dot Earth*,

[5] Revkin, A. C. (2014). In joint steps on emissions, China and US set aside "you first" approach on global warming. *The New York Times* (The Opinion Pages, *Dot Earth*), November 11. Retrieved from http://dotearth.blogs.nytimes.com/2014/11/11/in-joint-steps-on-emissions-china-and-u-s-set-aside-you-first-approach-on-global-warming/?_r=0http://dotearth.blogs.nytimes.com/2014/11/11/in-joint-steps-on-emissions-china-and-u-s-set-aside-you-first-approach-on-global-warming/?_r=0. While each *Dot Earth* entry is signed, they are not invited pieces by outsiders or by diverse staff writers (i.e. not really op-ed). Instead, Revkin is provided with an institutional space to make his own regular commentary, arguably, on behalf of the newspaper. It was thereby included as part of the sample of editorials and specific climate opinion sections.

a specialized online opinion section of the *New York Times* with an environmental focus. After some initial remarks, it provides the text of a joint statement made by the USA and China to outline their emissions agreement. The editorial ends with comments by David Victor, a senior climate policy analyst at the University of California, San Diego, who Revkin claims echoes his own feelings.[6] In his initial commentary, Revkin claims a "you first" rhetoric is what characterized deliberations in the past; that is an approach where countries refuse to act unless someone else acts first. This is an individualistic, unilateral tendency rather than a collectivistic one and thereby halts global action and social connection. The spatial aspect is addressed throughout the article with references to the USA and China as separate nations, but also as a team making a joint statement. Europe is also mentioned, and so is the whole world. It is said of the US–China joint statement that this "shift bodes well for the next round of negotiations". This expresses a desire modality (if something bodes well it is desired). Subsequently David Victor/Andrew Revkin states "[t]his is exactly what is needed" and thus expresses obligation regarding the agreement. "Collective action" is also presented as an obligation; it is "required". Collective action connects to solidarity, and "individual interests" are deemed not enough.

No references to redistribution, recognition, representation or any other specifics appear in the text. This makes it hard to determine what the "injustice" is. Hence, solidarity stands alone, without reference to the conceptions of justice that would motivate it. Justice does however appear implicitly in the clause "countries responsible for more than half the world's carbon dioxide emissions are accelerating their emission cutting plans". This indicates that responsibility is at work. It is however not specified (to the past or the future). Neither are the all-affected or the all-subjected principles.

"Immediate Global Action Needed to Cut Greenhouse Gas Emissions". **The Yomiuri Shimbun, *Japan*[7]**

This editorial's headline makes a call for global action that implies solidarity since everyone must get involved. The emphasis on "need"

---

[6]The portion of the article featuring the reprinted agreement is not analyzed here.

[7]Immediate global action needed to cut greenhouse gas emissions. (2014, November 6). *The Yomiuri Shimbun*. Retrieved from http://www.asianewsnet.net/ann_news.php?a=http://www.asianewsnet.net/Immediate-global-action-needed-to-cut-greenhouse-g&id=67155.http://www.asianewsnet.net/ann_news.php?a=http://www.asianewsnet.net/Immediate-global-action-

indicates obligation modality, which thus underlines the weight of the statement. The editorial fluctuates between past, present and future, as well as between global, international and national scales: "eighteenth century", "the 1990s", "2010", "late 2015", "2020", "2030", "2050", and "US", "China", "nations around the world", "international community", "global meetings". It calls for action in the form of reduced global emissions, pinpointing the USA, China and Japan. It states that the IPCC report "can be regarded as an effort to press each nation to take aggressive measures against global warming". To press someone evokes a desire modality, but the report is viewed only as an effort toward this and hence the probability of such an outcome cannot be considered high.

It is furthermore interesting how modality relates to spatial scales here. While all nations ("the international community"), and specifically the USA and China, "must" act globally, it is merely stated that it "would be advisable" for Japan to do the same. This is then modified further to be relevant only in the national context (where "it is imperative"), while internationally: "[f]rom the standpoint of a global emissions reduction, however, we believe it is even more helpful to use our nation's excellent energy-saving technology to help developing countries reduce their emissions". Thus, Japan allegedly serves best by teaching others to reduce. This resembles a statement in the headline of a secondary Swedish editorial: "Climate policy: We have the solutions".[8] Redistribution, recognition and representation are not in focus in the *Yomiuri Shimbun* editorial, and injustice is not considered an issue at all. It is rather solidarity alone that is advocated. The call for global action implies however the all-subjected principle, since everyone needs to act.

"Dire UN Climate Change Report is a Call to Action". Toronto Star, *Canada*[9]

This Canadian editorial stays relatively close to statements featured in the IPCC report or made by IPCC spokespersons. The "call to action" in the

needed-to-cut-greenhouse-g&id=67155.This is an un-signed piece termed "editorial" by the newspaper.

[8]Climate: We have the solutions. (2015, April 5). *Dagens Nyheter*. Retrieved from http://www.dn.se/ledare/huvudledare/vi-har-losningarna/http://www.dn.se/ledare/huvudledare/vi-har-losningarna/ This un-signed editorial is labeled "the main editorial" of the day.

[9]Dire UN climate change report is a call to action. (2014, April 1). *The Toronto Star*. Retrieved from http://www.thestar.com/opinion/editorials/2014/03/31/dire_un_climate_change_report_is_a_call_to_action_editorial.html.http://www.

headline refers to a quote from Rajendra Pachauri's speech at the release of the WGII report. There are brief references to the domestic context in Canada, partly made in a self-critical way: "Canadians might be forgiven for shrugging off worries about global warming [after a cold winter]. But that would be a mistake" and "[e]ven for Prime Minister Stephen Harper's climate-benighted government, this warning that some damage can't be reversed and that the worst is yet to come should be a wake-up call". The last sentence emphasizes obligation, something "should" be done. "Should" appears several times in the editorial. References to the IPCC report stress that its predictions "will" happen. Thus, obligation and truth/probability strengthen each other. An exception occurs when connections between climate change effects and poverty are discussed with the probability modality shrinking into "could". This resembles a secondary Swedish editorial piece, "Romson's good will does not help the poor"[10], where climate justice and justice concerning poverty are opposed, with poverty used as an excuse for not acting on climate change.

The Canadian editorial's calls for action connect to a solidaritarian mode. While remedies are explicitly mentioned, these are not remedies for injustice but rather technical remedies for climate change. By sticking closely to reporting at the IPCC press release event, and briefly situating this within the Canadian context, the editorial avoids touching upon what these calls for action might mean in terms of correcting wrongdoings and unjust conditions. The indigenous peoples of Canada are notably absent, despite a discourse about melting ices that would make such a connection quite obvious. The same tendency can be observed in editorial material from Sweden, a country where indigenous peoples live in the northernmost parts of the country.

Redistribution, recognition and representation are not discussed in the editorial, even though poverty and "tensions over water and other resources" are mentioned. The all-subjected principle however is implied in that Canadians must join others in action against climate change. Responsibility is thereby seen as collective, with spatial scales connected to both the global and the home country. Temporal scales of responsibility are not evoked.

thestar.com/opinion/editorials/2014/03/31/dire_un_climate_change_report_is_a_call_to_action_editorial.html.This un-signed editorial is featured under 'Opinion' and more specifically 'Editorials'.

[10]Lomborg, B. (2014). Romson's goodwill does not help the poor. (2014, November 2). *Svenska Dagbladet*. Retrieved from http://www.svd.se/romsons-goda-vilja-hjalper-inte-de-fattiga.http://www.svd.se/romsons-goda-vilja-hjalper-inte-de-fattiga. This is a signed piece under the heading 'Editorial'.

"Climate: Let's Move from Irresponsibility to Acts!" Le **Monde,** *France*[11]

This editorial mentions irresponsibility in the title and ends with a note on the same theme:

> In 2015, France will host the climate summit, where states will gather to sign an accord which would engage the entire planet, including long-industrialized countries historically *responsible* for global warming as well as new developing countries that are increasingly big polluters. It is a massive challenge in the face of which our country has to be a model (emphasis added).

Responsibility is termed historical, and thus related to restitution, which is an intrinsic part of a solidaritarian approach. This is combined with spatial scales: "the entire planet" should be engaged. The declaration that France should be exemplary, a model for others, sets this editorial apart from the others while simultaneously relating to them. On one hand, being "a model" by doing more sets the nation apart (rather than doing things indirectly through teaching, as in Japan and Sweden, or catching up with others, as in Canada). On the other hand, Japan and Sweden seem to position themselves as models for others, which informs their view of "teaching the others" rather than acting directly. In the French editorial, the action argument is underlined strongly by its obligation modality—France *must* be a model. "Sharing the burden" is evoked as well, resembling the Swedish editorial chronicle[12], and relating to the global spatial aspect (many nations share the burden). The article varies in its use of probability and obligation throughout. The verb "could" expresses probability, while "should not" and "must" express obligation. Probability occurs frequently while desirability ("let's hope for") is more rare. (Lack of)

---

[11] Climate: let's move from irresponsibility to action! (2013, September 29–30). *Le Monde*, p. 1. The editorial was translated by Armèle Cloteau, Rennes Institute of Political Studies, France. This is an un-signed editorial placed on the front page and labeled "Editorial".

[12] Persson, I. (2014). The climate threat—A political problem. *Aftonbladet*, November 3. Retrieved from http://www.aftonbladet.se/ledare/ledarkronika/ingvarpersson/article19794981.ab.http://www.aftonbladet.se/ledare/ledarkronika/ingvarpersson/article19794981.ab.For details, see the previous footnote on this article.

solidarity is for instance implied in the phrase: "So far they have offered us only words", which connects to the idea of solidarity as a brotherhood of action rather than mere phrases. The obligations include listening to calls for action by the USA, involving state leaders instead of environment ministers, and playing an exemplary role domestically.

While much of the emphasis is on action, the editorial does not provide any remedies for injustice. How should the unjust relationship between "historically responsible" countries and "new developed countries, big polluters" be tackled? By redistribution, political representation or what? This is not clarified. The all-subjected principle seems however to be implied by emphasis on the need for action globally *and* domestically.

CONCLUSIONS

The analysis suggests that solidarity was foregrounded in the argumentative discourse of editorial material published in high-income countries during the release of the IPCC reports. Calls for global action are common in the examined material, evoking an all-subjected principle rather than an all-affected one, asking everyone to take responsibility. In some articles, this is termed as responsibility for the past, in some for the future. Spatially, the *international* is foregrounded. This includes—or actually builds on the naturalization of—the *national* scale but excludes the *transnational* one. Thereby a *global* scale is not fully evoked, even though there are calls to global action. Justice is mentioned explicitly in the Swedish editorial chronicle, and often implied as a catalyst for discussions of who should take responsibility, but it is generally backgrounded and not specified in the examined articles. Remedies for climate change are occasionally mentioned (the Canadian editorial) but *remedies for injustice* are not. Even when the problem is stated as a political one (e.g. the Swedish editorial chronicle), it is not stated as an issue of political representation and left without solutions. Hence, while the solidarity needed for justice to be realized seems apparent, the nature of justice/injustice is not specified. Rather the discourse is limited to an abstract solidarity of words and washy evocations of the all-subjected principle.

This "washiness" is contrasted with a strong obligation modality underpinning the arguments in each of the examined editorial items. This is sometimes supported by further use of a negatively articulated permission modality, exemplified by claims that we must not shy away from responsibility for future generations (the Swedish editorial chronicle). Since the actions that must be taken are not specified (do they, for instance, have to do with redistribution, recognition or political representation?) it is ultimately *not the arguments, but rather the tone* that comes across as strong. Thus the arguments are still, in this sense, washy despite persistent calls to action.

It is also noteworthy that on some occasions (specifically in the Japanese editorial) the modality of the writing places a weaker demand on the home nation: other nations are obligated to act while action by the home nation is only desired or probable. This particular form of domestication (Olausson 2014; see also Chap. 4) puts less moral pressure on the home nation. A related form of domestication concurrently appears in terms of waiting for others to act before the home nation acts. This is implied in the Swedish editorial chronicle, and in the US article's reference to an era when the "you first" principle ruled. Yet another form of domestication is the hailing of the home country as a model for others (French and Japanese editorials, and a secondary Swedish editorial).

Since editorial material is generally skewed toward opinions that readers are assumed to like or agree with (Nord 1998, 2001), the outcome of this explorative study provides a glimpse into publicly acceptable and "correct" discourse in the examined high-income countries. This discourse is characterized by solidarity tendencies and a will to act, albeit not necessarily before others, as implied by evocations of the "you first" principle. This constitutes common ground, a joint history where everyone used the "you first" principle, with attitudes toward action now seeming to vary slightly. Things could thus be beginning to change. There are a few tendencies to abandon the "you first" principle. Take, for instance, the previously discussed French editorial urging the home country to act first or a secondary editorial from Japan expressing this tendency as a "no regrets" policy that "is now necessary for stopping global warming"[13], a call to action regardless of what others have done. Concurrently, the few explicit references to justice, and even fewer to remedies, indicate that dominant public understanding of climate change as a justice issue remains weak and unqualified. This expresses a readiness to act in solidarity but in a vague, non-specified way. The examined editorial discourse does not (yet) reassess economic privilege or the dynamics of political representation. It remains an emerging and only vaguely solidaritarian discourse.

[13] "No regrets" policy is now necessary for stopping global warming. (2013, October 3). *Asahi Shimbun*, p. 16. The editorial was translated by Shinichiro Asayama, Postdoctoral Fellow, National Institute for Environmental Studies (NIES), Japan.

BIBLIOGRAPHY

Askanius, T., & Uldam, J. (2012). Online social media for radical politics: Climate change activism on YouTube. *International Journal of Electronic Governance, 4*(1–2), 69–84.

Bullard, R. (1990). *Dumping in dixie: Race, class, and environmental quality.* San Francisco: Westview Press.

Caney, S. (2005). Cosmopolitan justice, responsibility and global climate change. *Leiden Journal of International Law, 18,* 747–775.

Chouliaraki, L. (2011). Improper distance: Towards a critical account of solidarity as irony. *International Journal of Cultural Studies, 4,* 363–381.

Chouliaraki, L. (2013). *The ironic spectator: Solidarity in the age of post-humanitarianism.* Cambridge, MA: Polity.

Couldry, N. (2012). *Media, society, world: Social theory and digital media practice.* Cambridge: Polity.

de Chavez, R., & Tauli-Corpuz, V. (2009). *Guide on climate change and indigenous peoples.* Baguio City, Phillipines: Tebtebba Foundation.

de Onís, K. (2012). Looking both ways: Metaphor and the rhetorical alignment of intersectional climate justice and reproductive justice concerns. *Environmental Communication, 6*(3), 308–327.

Deuze, M. (2005). What is journalism? Professional identity and ideology of journalists reconsidered. *Journalism, 6*(4), 442–464.

Eide, E. (2012). An editorial that shook the world: global solidarity vs. editorial autonomy. In E. Eide & R. Kunelius (Eds.), *Media meets climate: The global challenge for journalism* (pp. 125–143). Gothenburg: Nordicom.

Fenton, N. (2008). Mediating solidarity. *Global Media and Communication, 4*(1), 37–57.

Fowler, R. (1991). *Language in the news: Discourse and ideology in the press.* London: Routledge.

Fraser, N. (2000). Rethinking recognition: Overcoming displacement and reification in cultural politics. *New Left Review, 3,* 107–120.

Fraser, N. (2008). *Scales of justice: Reimagining political space in a globalizing world.* New York: Columbia University Press.

Fraser, N. (2014). Transnationalizing the public sphere: On the legitimacy and efficiency of public opinion in a post-Westphalian world. In K. Nash (Ed.), *Transnationalizing the public sphere* (pp. 8–42). Cambridge: Polity.

Fraser, N., & Honneth, A. (2003). *Redistribution or recognition: A political-philosophical exchange.* London: Verso.

Fraser, N., & Olson, K. (Eds.). (2008). *Adding insult to injury: Nancy Fraser debates her critics.* London: Verso.

Global Challenges Foundation. (2014, September 22). Fyra av fem anser att klimatförändringen hotar mäsnkligheten [Four out of five believe climate change threat-

ens humanity]. Retrieved from http://globalchallenges.org/2014/09/17/
fyra-av-fem-anser-att-klimatforandringen-hotar-manskligheten/
Kössler, R., & Melber, H. (2007). International civil society and the challenge for
global solidarity. *Development Dialogue, 49*, 29–39.
Kunelius, R. (2012). Varieties of realism: Durban editorials and the discursive
landscape of global climate politics. In E. Eide & R. Kunelius (Eds.), *Media
meets climate: The global challenge for journalism* (pp. 31–48). Gothenburg:
Nordicom.
Larsson, E. (2015). Svensk dagstidningsbevakning av klimatförändringar [Swedish
daily press coverage of climate change]. Bachelor thesis, Örebro University, Örebro.
Liedman, S. E. (1999). *Att se sig själv i andra: Om solidaritet [To see yourself in
others: On solidarity]*. Stockholm: Bonniers.
Nash, K. (Ed.). (2014). *Transnationalizing the public sphere*. Cambridge, UK:
Polity Press.
Nord, L. (1998). *Makten bakom orden: En studie av ledarsidor och ledarskrivande
i svensk dagspress [The power behind the words: A study of editorials and editorial
writing in the Swedish daily press]*. Stockholm: JMK.
Nord, L. (2001). *Vår tids ledare*. In *En studie av den svenska dagspressens politiska
opinionsbildning [The editorials of our time: A study of political opinion in the
Swedish daily press]*. Stockholm: Carlssons.
Nussbaum, M. C. (2013). Foreword. In I. M. Young (Ed.), *Responsibility for jus-
tice* (pp. ix–xxv). Oxford: Oxford University Press.
Olausson, U. (2014). The diversified nature of "domesticated" news discourse:
The case of climate change in national news media. *Journalism Studies, 15*(6),
711–725.
Owen, D. (2014). Dilemmas of inclusion: The all-effected principle, the all-
subjected principle, and transnational public spheres. In K. Nash (Ed.),
Transnationalizing the public sphere (pp. 112–128). Cambridge: Polity.
Page, E. (2006). *Climate change, justice and future generations*. Cheltenham, UK:
Edward Elgar.
Rao, S., & Wasserman, H. (Eds.). (2015). *Media ethics and justice in the age of
globalization*. Hampshire: Palgrave Macmillan.
Rawls, J. (1971). *A theory of justice*. Harvard: Harvard University Press.
Rhaman, M. (2014). Justice delayed-justice denied: Framing climate justice in
Bangladeshi newspaper journalism during COPs. In B. León (Ed.),
Communicating climate change: From global agenda to media representation
(pp. 179–212). Salamanca: Comunicación Social.
Roosvall, A. (2014a). Beyond dialogue: Exploring solidarity as a mode of com-
munication through a debate on readers' comments to online news. *Northern
Lights: Film and Media Studies Yearbook, 12*(1), 49–67.
Roosvall, A. (2014b). The identity politics of world news: Oneness, particularity,
identity and status in online slideshows. *International Journal of Cultural
Studies, 17*(1), 55–74.

Roosvall, A., & Tegelberg, M. (2012). Misframing the messenger: Scales of justice, traditional ecological knowledge and media coverage of indigenous peoples and climate change. In E. Eide & R. Kunelius (Eds.), *Media meets climate: The global challenge for journalism* (pp. 297–312). Gothenburg: Nordicom.

Roosvall, A., & Tegelberg, M. (2015). Media and the geographies of climate justice: Indigenous peoples, nature and the geopolitics of climate change. *TripleC: Communication, Capitalism & Critique, 13*(1), 39–54.

Segerberg, A., & Bennett, W. L. (2011). Social media and the organization of collective action: Using Twitter to explore the ecologies of two climate change protests. *Communication Review, 14*(3), 197–215.

Sen, A. (2009). *The idea of justice.* Harvard: The Belknap Press of Harvard University Press.

Shepard, P., & Corbin-Mark, C. (2009). Climate justice. *Environmental Justice, 2*(4), 163–166.

Shue, H. (2014). *Climate justice: Vulnerability and protection.* Oxford: Oxford University Press.

Silverstone, R. (2007). *Media and morality: On the rise of the mediapolis.* Cambridge, UK: Polity.

Smiley, M. (2010). Collective responsibility. On Stanford Encyclopedia of Philosophy web page of Stanford University Press. Retrieved from http://plato.stanford.edu/entries/collective-responsibility/

Uldam, J., & Askanius, T. (2013). Online civic cultures: Debating climate change activism on YouTube. *International Journal of Communication, 7,* 1185–1204.

Young, I. M. (2013). *Responsibility for justice.* Oxford: Oxford University Press.

Ytterstad, A., & Russell, A. (2012). Pessimism of the intellect and optimism of the will: A Gramscian analysis of climate justice in summit coverage. In E. Eide & R. Kunelius (Eds.), *Media meets climate: The global challenge for journalism* (pp. 247–262). Gothenburg: Nordicom.

CHAPTER 7

Emerging Economies and BRICS Climate Policy: The Justifying Role of Media

Dmitry Yagodin, Débora Medeiros, Li Ji, and Ibrahim Saleh

On April 22, 2015, environment ministers from each of the BRICS countries gathered in Moscow to discuss common economic development issues in the context of climate change. This was their first official meeting with such an agenda. During the meeting, the BRICS ministers discussed plans to cooperate on issues of environmental protection. As a means of addressing the issue of climate change, they agreed to establish a "platform of public-private partnership for exchanging the best practices available and promoting green technology" (BRICS 2015). Securing "economic

D. Yagodin (✉)
School of Communication, Media and Theatre, University of Tampere, Tampere, Finland

D. Medeiros
Institute for Communication and Media Studies, Free University of Berlin, Berlin, Germany

Li Ji
School of Journalism and Communication, Wuhan University, Wuhan, China

I. Saleh
Political Communication Department, Cairo University, Cairo, Egypt

© The Author(s) 2017   151
R. Kunelius et al. (eds.), *Media and Global Climate Knowledge*,
DOI 10.1057/978-1-137-52321-1_7

growth" in changing natural environments was identified as a priority. Yet none of the publicly available transcripts of statements from the meeting mention the IPCC AR5 or its warnings about the economic costs of climate change and the need to reduce consumption and emissions globally.

This chapter looks at four of the BRICS countries[1] as a useful framework to study the role of the media in justifying environmental policies in light of the IPCC report. It also looks at the usefulness of applying the BRICS concept to the contexts of economic growth and climate change. The country choice is significant for two reasons. The first reason relates to power. The alliance of "BRICS-countries" is a product of economic development. Brazil, Russia, India and China began using BRIC to refer to their alliance, after the acronym was invented in a 2001 report by global investment banking firm Goldman Sachs (O'Neill 2001). The four countries started formally convening for summits in 2009 with South Africa joining the group in 2011. In research literature and public discourse, the BRICS countries share the status of "emerging economies", though in the historical trajectories of Brazil, South Africa and, especially, Russia this is perhaps characterized better as a "come-back" than an initial emergence (Becker 2013: p. 5). The BRICS group is also sometimes perceived to be collaborating in the pursuit of an alternative global order (de Coning et al. 2014) or associated with the rise of new cultures of communication (Straubhaar 2015).

The second reason for this choice of focus is environmental. The BRICS countries are crucial global players in coping with climate change. Without their efforts to reduce emissions "climate change mitigation will prove very difficult" (Leal-Arcas 2013: p. 22). In 2007, China surpassed the USA as the highest global emitter of carbon dioxide (CO^2). Endowed with vast territories, the BRICS countries are together responsible for approximately one-third of all greenhouse gas emissions in the world, a figure that includes emissions caused by deforestation (Clark 2011). The growing climate responsibility of the emerging economies in the BRICS group, and China in particular, leads to important policy negotiations and public discussions in the respective countries and between them.

Therefore, the climate strategies that are publicly discussed in the BRICS countries constitute the basis for future climate negotiations and geopolitical positioning. A key question then becomes: how does the media in BRICS countries represent and justify existing and potential national

---

[1]We did not include India in the analysis for reasons of consistency with the broader research context presented in this volume.

climate policies with regard to the recent IPCC findings? To answer this question, the chapter considers two factors: the realities of the existing national media systems and the natural urge to minimize any threats to national economic growth. Thus, the chapter begins with an overview of these media systems. It continues with a brief summary of the key socio-economic factors and climate change realities in the selected countries. The subsequent empirical analysis focuses on media coverage of the IPCC reports (content analysis of the *voices* mentioned in the media texts and comparison of thematic *representations* across the selected countries).

NON-WESTERN MEDIA SYSTEMS

Hallin and Mancini (2004) argue that many non-Western media systems fit their "polarized pluralist" model, which they differentiate from "democratic corporatist" and "liberal" models. This claim, however, groups a wide range of dissimilar media systems, placing them together with media systems dominant in Southern Europe—the original context for the polarized pluralist model (low press circulation, high political parallelism between the media and political parties, weak journalistic professionalization and strong state intervention). The oversimplified categorization has been contested and corrected in application to other national contexts (Hallin and Mancini 2012) and to the BRICS countries in particular (Nordenstreng and Thussu 2015).

First, take the case of Brazil. The Brazilian media system cannot be explained by the notion of political parallelism (de Albuquerque 2012: p. 73). In Brazil, party distinctions are not emphasized on the national political and public agendas and mass media tends to view its audience as encompassing the whole nation. The leading Brazilian newspapers have relatively low circulation (television is the main medium), with urban elites as their primary audience. Though the role of state subsidies is high, the share of state ownership and intervention is minimal (de Albuquerque 2012: pp. 78–86) and the majority of the media companies in Brazil are private family enterprises. *O Globo* newspaper, with its head office in Rio de Janeiro, is part of *Globo Organizations*, a media conglomerate controlled by the Marinho family. *O Globo* is one of the two Brazilian newspapers analyzed here. The second, *O Estado de S. Paulo*, which belongs to the Mesquita family, is part of the *Grupo Estado* conglomerate. The newspaper is politically conservative but adheres to liberal economic values, defending privatization and advocating for reduced state regulation of the market. *O Globo* and *O Estado de S. Paulo* are among the most influential elite newspapers in the country.

Grouping the Russian media system together with other Eastern European contexts—where elements of "state paternalism" and "political clientelism" (Jakubowicz 2007: p. 304) resemble the model of polarized pluralism—is also insufficient (Vartanova 2012: p. 122). Interdependence between certain media and across different political parties (i.e. high political parallelism) is uncommon in Russia. Understanding the Russian system requires more nuanced considerations of the role of the state-controlled quality press in intra-elite horizontal communication (Roudakova 2008: p. 44). As Vartanova (2012: p. 134) points out, maintenance of the vertical channels of communication, however, has been the task of national television. She suggests describing the Russian media system in terms of a "statist commercialized model" that combines economic liberties with mostly informal and indirect control over media content. Such a media system splits into two unequal parts: the mainstream part predominantly made up of state-controlled press and television; and a smaller part made up of independent news sources. The tightening of political freedoms since the 2000s has forced this smaller (and politically engaged) segment to either "immigrate" to online forums (Vartanova 2012: p. 142) or to leave the country and settle in neighboring Ukraine or the Baltic states. The two Russian newspapers included in this study, *Rossiyskaya Gazeta* (RG) and *Kommersant*, belong to the larger mainstream part of the media system. It is also a common trend that such elite newspapers have low readership but exhibit high citation ratings among all national news media. *RG* is a governmental daily that covers a broad spectrum of news, opinions and official decrees. The privately owned *Kommersant* is part of the business empire of one of the richest Russian oligarchs, Alisher Usmanov. *Kommersant* focuses on business, economics, foreign and domestic politics.

In order to depict the Chinese media system, and compare it to others, one must draw a complex picture where Leninist and Maoist ideological backgrounds intersect with "ongoing struggles between different universalisms and different regimes of truth" (Zhao 2012: p. 150). The role of the state and the Communist party is overwhelming. The state owns virtually all major mass media. Similar to Russia, the Chinese leadership recognizes the importance of free market mechanisms and profit-making, but exerts more formal institutional control and direct intervention at the level of media content. There is some political media pluralism in China but it only reflects subtle intra-party divisions and tensions (Zhao 2012: p. 163); another commonality between the Chinese and Russian media systems. Although television is an important tool for mass communication across the large nation, the press is still considered crucial for reaching the public and not only elite groups. *People's Daily*, for example, is the central organ of the Communist party, whose official vision of the press' role states that "the party's mouthpiece is also the mouthpiece of the 'people'" (Zhao 2012: p. 164). *China Youth Daily*, another newspaper considered here, is published by the Communist Youth League of China. Thus, both newspapers unequivocally reflect state policies and, unlike the quality press in Brazil, Russia and South Africa (see below), address much wider audiences than just narrow elite groups.

What distinguishes the South African media system from the others described here is that it caters to a highly heterogeneous (racially, linguistically, ethnically) society and combines characteristics from each of Hallin and Mancini's three models (Hadland 2012). The English-language press, in particular, is the product of British influence (the liberal model) with similar editorial traditions, commercialism and a commitment to free press values. However, newspapers in Afrikaans and other languages traditionally display more partisan and advocacy journalism with elements of clientelism more characteristic of the polarized pluralism model. In South Africa, there is a strong emphasis on national development issues and hence a tendency for journalists to focus on national interests rather than on public interests (Hadland 2012: p. 107). The state's role has also been increasingly visible in the South African media system. In spite of the existing constitutional rights, cases of restricted access to certain (critical) information and some official statistics are common, along with other forms of government intervention. In the overall structure of the South African media market, the quality English-language press occupies a small yet important elitist niche of low readership. *Business Day* and *Cape Times* are two elite dailies based respectively in Johannesburg and Cape Town. *Business Day* is a nationally distributed daily that focuses on economics. *Cape Times* is a regional source of general current affairs news.

SOCIO-ECONOMIC FACTORS

The BRICS concept conceals many differences under its homogenizing mask (Sparks 2014). A broader socio-economic context can help to clarify the distinctions. They also provide an important background to explore how BRICS news media received the AR5 and used it to discuss economic issues and justify possible climate policies. We wish to situate them in relation to the *contexts* in which the media coverage was produced in order to provide background for the comparison of actual *representations* in media *texts*. The theoretical models described above can be further substantiated with empirical contexts of socio-economic development.

Table 7.1 lists seven factors locating Brazil, China, Russia and South Africa against each other and, as a point of reference and comparison, against similar indicators in the USA. In our four target countries, the only comparable indicator is human development index (HDI), which accounts for life expectancy, education levels and standards of living. The Media Freedom index (MFI) shows drastic differences, with China the least free, though the Internet penetration (Web) gap is less profound. Public perception (PRP) of climate-related risks is clearly higher in Brazil, whereas others are comparable to each other and lower than the USA. However, people's perceptions do not always directly reflect the actual risks a nation faces. The Climate Risk index (CRI) measures the number of casualties and economic damages caused by extreme weather events over the last 20 years (Germanwatch 2015) by assigning lower scores to higher risks. China, Russia and the USA have similarly high risks, whereas Brazil and South Africa are less affected. The discrepancy in levels

Table 7.1 Socio-economic factors in 'emerging economies' and the USA

	GDP[a]	CO2[b]	MFI[c]	Web[d]	HDI[e]	CRI[f]	VAI[g]	PA[h]	PRP[i]
Brazil	11.89	2.56	31.9	51.6	0.74	85.7	54.1	77	96.4
China	6.96	7.42	73.6	45.8	0.72	44.7	58.6	75.6	60.4
Russia	14.47	12.63	45	61.4	0.78	43.3	65.9	83.9	56
South Africa	6.89	6.25	22.1	48.9	0.66	83.8	53.9	66.2	69.9
USA	52.94	16.55	24.4	84.2	0.91	44.7	76.5	98.4	79.7

[a]GDP per capita in thousands of USD (International Monetary Fund 2013; Statistics Norway 2013)

[b]Carbon emissions per capita in tonnes (EDGAR 2013)

[c]Media freedom index where the lowest score indicates more freedom (Reporters Without Borders 2015)

[d]Percentage of individuals using the Internet (ITU 2013)

[e]HDI estimates life expectancy and living standards, where the highest score represents better results (UN Development Programme 2013)

[f]Climate risk index for 1994–2013 where the lowest score means the highest risk (Germanwatch 2015)

[g]Vulnerability and adaptation index (ND-GAIN 2013), higher score indicates better preparedness to resist threats of climate change

[h]Public climate change awareness by percentage (Lee et al. 2015, data from Gallup World Poll 2007, 2008)

[i]Public climate change risk perception (Lee et al. 2015, data from Gallup World Poll 2007, 2008). Percentage of those indicating "serious" personal risk as a share of those deemed "aware"

of public perception and the actual results of risk assessments raise crucial questions for the study of science communication.

Another notable distinction is the level of carbon emissions per capita (CO^2) that places Russia far ahead of its "emerging" counterparts. The lowest level of CO^2 in Brazil does not prevent the country from having a relatively high GDP, expressing the moderate contribution of the energy industry to the national economy. China and Russia are radically different in terms of climate risk. South Africa, while having the lowest GDP and HDI, maintains the highest level of media freedom within the group.

Brazil, Russia, China and South Africa also have different domestic aspirations with regard to climate policy. One example that stands out is Russia's absence from a climate policy alliance known as BASIC that includes Brazil, India, China and South Africa. At previous climate summits, these nations have been resolute in acting jointly in global climate policy negotiations. An important milestone in this concerted action occurred at the 2009 summit in Copenhagen (COP15) where BASIC, mostly lead by China and India, united their bargaining power to block the idea of equal mitigation efforts for developed and developing countries (Wu 2012). Russia's passive role in these negotiations stems from its comfortable position as an observer, with relatively low mitigation obligations (a result of the 1997 Kyoto protocol), that can afford to keep "waiting for the best deal" (Andonova and Alexieva 2012: p. 623). Since 2004, when Russia finally joined the Kyoto protocol under terms favorable to the country's economic development, overall public interest in climate change has remained low (Rowe 2012; Poberezhskaya 2015).

In contrast, South Africa was among the key nations participating in the final stages of bargaining at COP15 and hosted COP17 in 2011. This symbolic advancement toward the center of global climate negotiations caused mixed feelings inside the country. Particularly during the COP17 summit in Durban, where the discourse of national pride in South African media coincided with more critical discussions about the costs of playing a leading role in climate change negotiations (Saleh 2012: p. 61). However, generally speaking, coverage of environmental issues has not been prominent in the South African daily press (Schreiner and Bosman 2012), and it was only recently, for instance, that *Business Day* appointed reporters specializing in environmental news.

In countries with high racial and social segregation (Brazil and South Africa), climate science reporting can pose challenges for journalists. In South Africa, it is not easy to convince people to save electricity when they do not even have access to it (Orgeret and d'Essen 2012: p. 271). Moreover, Orgeret and d'Essen refer to a *Business Day* reporter who admits that due to a general lack of science expertise among South African journalists it is difficult to comprehend what scientists say about complex climate problems. In China, journalists may feel they have been forced to frame many global news stories as conflicts. For example, fighting against negative images of China in climate change reporting has become a common obligation of Chinese reporters (Xu 2010).

These contexts are essential for understanding media interpretations of climate politics and the IPCC reports. The interpretations, in their turn, shape public discourses where different social actors (media and media sources) occupy and express particular positions.

KEY INTERPRETERS OF THE IPCC REPORT AND CLIMATE CHANGE

How, then, did newspapers from these different media traditions and national contexts represent the AR5? What actors were quoted as sources of expert knowledge and opinion? What arguments were deemed constitutive of the coverage? The "voices" from the national samples are crucial for linking the contexts outlined above to subsequent analysis of interpretations.

Table 7.2 shows the distribution of voices in the national AR5 samples.[2] The key distinctions here highlight domestication patterns. Brazil's 105 newspaper articles provide a rich variety of materials containing 167 quoted voices, roughly 1.59 voices per story. The sample from South Africa has only 26 stories, but 52 voices or 2 voices per story, suggesting (at story level) a more dialogic journalistic culture with less space dedicated to short, "voiceless" news. The Russian and Chinese samples of 11 and 5 stories respectively contain 19 (1.72 per story) and 5 (1 per story) voices.

Quite naturally, the analyzed media represented the IPCC report predominantly through the voices of science. A rough estimation of voice composition across the samples shows that science, both domestic and foreign, account for between 45 and 70 percent of all voices, depending on the country. We can see the highest domestication patterns in Russia

---

[2] This includes stories that feature the IPCC AR5 as the main topic and general climate change stories that refer to the IPCC AR5.

Table 7.2 Voice categories and comparison to the global AR5 sample (percent)

	Brazil (N = 167)	Russia (N = 19)	China (N = 5)	S. Africa (N = 52)	Global average
Transnational politics	6	5.3		9.6	11.4
Domestic actors	42.4	63.2	60	53.8	52.1
– politics	3			11.5	10.9
– science	29.9	36.8	60	30.8	28.6
– civil society	5.4	26.3		7.7	10.7
– business	1.2				1.9
Foreign actors	49.1	31.6	40	34.6	33.1
– politics	4.2	5.3	20	9.6	6.3
– science	34.7	26.3	20	15.4	19.2
– civil society	10.2			7.7	6.2
– business	1.8			5.8	1.4
Media	3.6			1.9	3.1

and South Africa. China's coverage is too small for comparison, whereas Brazil is a convincing case of significant reliance on foreign categories of voices with foreign scientists as the largest contributing group.

Table 7.3 draws a more detailed picture by pairing the media included in the study with the respective number of AR5 stories for each newspaper. The table also lists the names of journalists or other external authors, like Zheng Guoguang from China Meteorological Administration, with the number of stories they published. The list is limited to the most prolific authors in their respective countries. Finally, the table shows the most frequently quoted voices and the number of stories they were featured in.

References to voices (shown in bold) that cut across national samples are often self-evident (Rajendra Pachauri, Ban Ki-moon, Thomas Stocker [Working Group I (WGI)] and Chris Field [WGII]), since they represent the leadership of the UN and the IPCC. The Brazilian coverage in particular included the voices of many IPCC-related officials and scientists. Brazilian domestic science, for example, was represented by Jose Marengo. In other countries, there was no such pronounced dependence on this group of voices. Another notable finding is the direct reliance on the journalistic output of *Reuters'* Alister Doyle in Brazil and South Africa, a sign of a global journalistic field with an increasing dependency on news agencies. The following inquiry into the content of the coverage shows how some of the voices are used in national media discourses and what thematic categories feature prominently in the coverage.

Table 7.3 Summary of the media samples, authors and voices

Brazil	Russia	China	South Africa
Media samples and number of relevant stories			
O Estado de S. Paulo (76)	Kommersant (5)	China Youth Daily (1)	Business Day (15)
O Globo (29)	Rossiyskaya Gazeta (6)	People's Daily (4)	Cape Times (11)
Names of authors and number of their stories			
Andrei Netto (18)	Yuri Medvedev (3)	Edward David (1)	Melanie Gosling (4)
Denise Chrispim Marin (10)	Angelina Davydova (2)	Liu Zhonghua (1)	**Alister Doyle** (4)
Renato Grandelle (9)		Tian Hong and Li Yingqi (1)	Philip Lloyd (2)
Giovana Girardi (8)		Zheng Guoguang (1)	Sue Blaine (2)
Alister Doyle (7)		Zheng Hong (1)	
Names and number voices quoted			
Rajendra Pachauri (15)	Aleksey Kokorin (3)	Qin Dahe (1)	Bob Scholes (4)
Suzana Kahn (9)	**Rajendra Pachauri** (1)	Chen Deliang (1)	Debra Roberts (3)
Jose Marengo (8)		Edward David (1)	Philip Lloyd (3)
Thomas Stocker (8)	**Thomas Stocker** (1)	Zheng Yan (1)	**Rajendra Pachauri** (2)
JeanPascal van Ypersele (5)	**Chris Field** (1)		Bruce Hewitson (2)
Carlos Rittl (4)	Aleksandr Nahutin (1)		Crispian Olver (2)
Ban Ki-moon (3)	Barack Obama (1)		Jane Olwoch (2)
Carlos Nobre (3)	Mototaka Nakamura (1)		John Kerry (1)
Chris Field (3)	Olga Solomina (1)		Kaisa Kosonen (1)
Christiana Figueres (3)	Poma Pomak (1)		Ban Ki-moon (1)
	Richard Harrison (1)		

THEMATIC ANALYSIS OF THE MEDIA CONTENT

For thematic analysis we selected eight stories, two from each country, that either explicitly focus on the IPCC (preferably referring to WGII and WGIII) or discuss national climate policies with references to the IPCC findings. The examples are stories with a high rate of attention measured by story size. They focus on the economy or national adaptation and mitigation strategies. We preferred opinionated genres over hard news. This section introduces and discusses these materials in four country-based subsections.

Biofuel and Responsibilities in Brazil

The WGII results received vivid coverage in Brazil's *O Estado de S. Paulo*. First, Giovana Girardi reported on an alleged controversy with the headline "Biofuels are left out of the IPCC report before discussion".[3] She wrote that the Brazilian delegation requested the inclusion of a passage in the report highlighting the negative consequences of biofuels. However, the passage was crossed out before it was even debated.

The story reminds readers that biofuel is central to the Brazilian economy and that it produces less emissions than fossil fuels. However, it points out that the expansion of biofuel use threatens the biodiversity of several Brazilian regions. The government's position in the IPCC negotiations comes as no surprise. Officials in the stories try to emphasize the positive side of biofuels and downplay the negative side effects (such as the expansion of sugarcane production into vulnerable areas like the Amazon). This position is articulated by Jose Miguez, a representative of the Ministry of Environment, who says "Brazil will always defend biofuels because they not only mitigate emissions, but also create jobs and development". According to the story, the controversy stems from the fact that the IPCC originally relied on biofuel research conducted by Brazilian scientist David Lapola. Apparently the IPCC authors misunderstood his study and hence drew false conclusions about the threats to nature. The article settles the debate with Lapola himself suggesting "the truth lies between these two extremes", that is, between the positions of the government and the IPCC. The overall critical tone of the article thus targets neither the government nor the science (IPCC). Rather it is a critique of the *intergovernmental negotiation* process in which the decision to exclude the controversial biofuel clause from the final draft of the report was not properly explained to the public.

Science journalist Herton Escobar sets out a broader context for domestic criticism in an interview titled "Our economic model is the villain of the environmental crisis".[4] The main argument comes through the

---

[3] Girardi, G. (2014). Biofuels are left out of the IPCC report before discussion. *O Estado de S. Paulo*, March 30. Retrieved from http://sustentabilidade.estadao.com.br/noticias/geral,biocombustiveis-sao-cortados-de-relatorio-do-ipcc-antes-de-discussaohttp://sustentabilidade.estadao.com.br/noticias/geral,biocombustiveis-sao-cortados-de-relatorio-do-ipcc-antes-de-discussao,1147112

[4] Escobar, H. (2014). Our economic model is the villain of the environmental crisis. *O Estado de S. Paulo*, April 7. Retrieved from http://sustentabilidade.estadao.com.br/noticias/geral,nosso-modelo-economico-e-o-vilao-da-crise-ambientalhttp://sustentabilidade.estadao.com.br/noticias/geral,nosso-modelo-economico-e-o-vilao-da-crise-ambiental,1150634

voice of an interviewee, Eduardo Felipe Matias, a law expert and author of a new book on sustainable development, *A Humanidade contra as cordas* (Humanity against the ropes). In the interview, Matias criticizes Brazil's hesitance to pursue a more sustainable path and calls for more investment in biofuels. This opinion aligns with the *sustainable development discourse* that we find at the state level (d'Essen 2010: p. 85). Matias simultaneously defends a multi-stakeholder approach by pointing to the fact that the role of individual actors and businesses can be more prominent. Not fully consistent with neo-liberal values, however, he also states that the current capitalist model, based on consumption, needs serious change: "If you stop to think about who the bad guy is in this story, it is our whole economic model as a society". The argument is that companies and countries must take a sharp turn toward sustainable development to have a competitive edge. He adds that every individual has a responsibility "as a part of this system in which we are inserted, be it as voters, consumers or simply as citizens. [...] There will be no silver bullet and it won't be one big player or one big treaty that will solve the problem".

In this second Brazilian example, we see the introduction of *system-critical discourse*. The media delegates the right of critical evaluation to an external moral authority, providing a public platform through the interview genre. Matias's book provides a pretext for the journalist to take a story angle that function as both a promotion for the book and a link guiding the audience toward a more philosophical conception of this topic.

Energy Efficiency Through Natural Gas in Russia

In a scarce Russian sample, the two selected stories are the product of an individual effort rather than a systematic editorial policy. Like in Brazil, the two stories come from the same newspaper, namely *Kommersant*. Written by educator, environmental campaigner and journalist Angelina Davydova, both stories only appeared online and emphasized the economic side of the climate science.

The first story, "Climate change price is growing"[5], reflects on WGII. The author reiterates IPCC findings on the universal impacts of climate change effects on economies throughout the world. The "outdated" claim that climate change will cause damage costing up to 3.6 percent of the global

---

[5] Davydova, A. (2014). Climate change price is growing. *Kommersant*, April 1. Retrieved from http://www.kommersant.ru/doc/2443235http://www.kommersant.ru/doc/2443235

GDP (see e.g. Ackerman and Stanton 2008), is corrected in a quote from WGII co-chair Chris Field who suggests taking into account the potential multiplication of risks and hence a significantly higher price. In the text, the IPCC authority receives additional legitimation through references to a similar report by the Russian meteorological agency, *Roshydromet*. Whereas the IPCC is said to predict that most climate change damage will be caused in African and Asian countries, national authorities are quoted to stress that current "temperature deviations are much higher [in Russia] than in the world on average". This domestic voice, however, is represented by a statement from the Russian Head of the WWF Climate and Energy Program, Aleksei Kokorin. The story draws on the IPCC report to state the necessity of a systemic approach to risk management and climate investing. As the author concludes, "it is precisely the isolated management of separate industries […] that has largely been the cause of current ecological problems". This passage resonates with the discussion of sustainable economic development observed in the Brazilian example above.

Davydova's second story, "Ecologists rehabilitated the gas"[6], is a response to WGIII. The story cites Aleksei Kokorin, once again, to articulate domestic solutions for "structural and technological modernization" that could help increase the energy efficiency of Russia's economy by 40 percent by 2020. While the main purpose of WGIII (mentioned in the story) was to elevate the role of renewable energy, Russian domestic discourse is more enthusiastic about gas being considered cleaner than most other traditional sources of energy. As the story subtitle suggests, "UN experts invite investments into gas industry for the sake of climate". This, according to the newspaper, corresponds with Russia's interest in improving energy efficiency. Thus, the call for investment into a "green economy" is rearticulated into specific national interests closely tied to the gas industry (*domestic green economy* discourse). This is further justified by a quote from Janusz Bialek, a researcher at Durham University, who says "gas stations, due to their capacity to quickly adjust production volumes to changing demands, are able to create the conditions necessary for developing wind and solar stations in many countries".

References to Kokorin, as a representative of the WWF, do not discuss his role as a civil society actor; rather he is introduced as a member of the expert community. This coincides with a recent trend in Russian media and

[6]Davydova, A. (2014). Ecologists rehabilitated the gas. *Kommersant*, April 18. Retrieved from http://www.kommersant.ru/doc/2456494http://www.kommersant.ru/doc/2456494

governmental policy to either co-opt (selective sponsoring of loyal organizations) or marginalize international NGOs (sometimes referred to as "foreign agents"). Allowing civil society "experts" to occupy the role of climate change voices in the Russian public domain, as well as an absence of mid-level authorities responsible for climate and energy policies, can be viewed as evidence of the low priority that the state attributes to environmental issues (Andonova and Alexieva 2012). Perhaps this also explains why the two most relevant Russian stories only appeared online, in media space more suitable for alternative voices and politically controversial topics.

National Strategy and Global Obligations in China

The Chinese newspapers included in this analysis emphasize the discourse of anthropogenic causes of climate change. From the coverage it follows that change is unavoidable and that China is resolute to face the challenges and responsibilities as a developing and key emerging economy. In the two newspapers studied, there was no consistent interest in the subject of the IPCC reports, nor did any of the journalists publish more than one text about it.

The prominent voice in *People's Daily* is a domestic national political actor, the head of China's Meteorological Administration, Zheng Guoguang. He is the author of a long commentary piece titled "Implement the national strategy for adapting to climate change".[7] In the text, Zheng calls for cooperation at all levels of Chinese government along the path of implementing the strategy. He confirms the certainty of the IPCC scientific conclusions regarding the human causes of climate change, and points out that "the IPCC's WGII report also provides many recommendations on preferable practices, skills and approaches for China to deal with climate change, prevent disasters and build eco-civilization". Based on the fundamental recognition of the IPCC results, this example of Chinese media discourse emphasizes full resolution. It emphasizes the aim of developing a strategy for adapting to climate change according to national interests, and thereby as a "contribution to the lasting development and happiness of our future generations". Similar to the focus in one of the Russian stories, this Chinese commentary mentions the economic costs of climate change (up to 2 percent of the global GDP).

The lack of voice diversity in the coverage and the overall uncontested tone of the media discourse in China is a sign that climate change is not a

---

[7]Guoguang, Z. (2014). Implement the national strategy for adapting to climate change. *People's Daily*, April 15. Retrieved from http://cpc.people.com.cn/n/2014/0415/c64102-24895794.html

debatable or negotiable issue. The few voices that are present in the texts indicate the high significance of Chinese domestic political actors and a conviction that any adaptation plans will be treated as internal political issues. As a party-owned newspaper, *People's Daily* strictly reflects the state climate change policy. It sends a clear signal to its readers that government decisions are thoroughly prepared, based on scientific proofs, and that they should be unconditionally followed. Being part of the traditional discourse of *national harmony and stability* (Liu 2011), Chinese media underpin the authority of the central government.

In November 2014, after the IPCC synthesis report was presented, much of the media coverage overlapped with a major climate deal between China and the USA. Predictably, *People's Daily* reported extensively on this Joint Announcement. This time, stories that mentioned the IPCC or climate change included high-level domestic and foreign political actors (Xi Jinping and Barack Obama). The reporting describes the agreement between the two leading economic powers to jointly promote the upcoming UN climate negotiations in Paris and to strengthen cooperation in areas of clean energy and environmental protection. In contrast to the April reporting, at this later stage of media coverage, climate change is treated as an international issue due to a number of top-level deliberations. For instance, after the 2014 Asia-Pacific Economic Cooperation meeting, a *People's Daily* news story titled "The responsibility of the great country can light a dream"[8] quotes a climate expert praising China's determination to reduce carbon emissions.

Without visible debates or controversies in the relatively scarce media coverage of the IPCC it is still clear that the Chinese government faces international pressure as the country with the highest level of CO^2 emissions. Moreover, China evidently remains committed to achieving its mitigation goals and improving its global reputation, even if it is not good timing for its economic development.

Policy Efficiency in South Africa

Reacting to the WGI report, South Africa's *Business Day* opens its coverage with a story echoing problems raised in Brazil and Russia regarding alternative energy sources. Journalist Sue Blaine[9] urges the government

[8] *People's Daily* (2014). The responsibility of the great country can light a dream. *People's Daily*, November 14. Retrieved from http://cpc.people.com.cn/n/2014/1114/c64387-26022467.html\http://cpc.people.com.cn/n/2014/1114/c64387-26022467.html\
[9] Blaine, S. (2013). SA needs to find innovative ways to reduce cost of renewable energy sources. *Business Day*, September 27. Retrieved from http://www.bdlive.co.za/national/

to "look at innovative ways of reducing the financial costs of renewable energy sources", paraphrasing a statement by Saliem Fakir, Head of WWF's Living Planet Unit in South Africa. Fakir is the key civil society voice that the article mobilizes to criticize the South African government for not paying enough attention to policy initiatives that could "reduce outstanding barriers to providing renewable energy". To demonstrate the scope of the climatic changes that these policies must address, the article quotes WGI co-chair Qin Dahe as evidence of a global perspective and uses an interview with South African meteorologist Mxolisi Shongwe to provide a domestic angle.

Melanie Gosling, the most prominent South African climate reporter in our sample, was most productive during the publication of WGII and WGIII. Her report "Swift action vital to avoid disaster"[10] draws on voices of domestic climate science and local environmental authorities. The main voice is South African climate scientist and IPCC author Bruce Hewitson. His projections draw a highly domesticated picture that addresses citizens directly ("imagine a summer hotter than any you have experienced in South Africa, and waking up in the morning to no shower because you have used your monthly quota of water"). To avoid such a scenario, the scientist calls for immediate policy changes and puts responsibility on the national government: "I would like to invite every South African minister to dinner and say: 'Give me three hours of your time'".

To address this problem on a more practical level the story points to the need for a decarbonization of the South African economy. Gosling builds her argument by using the words of Bob Scholes, an ecologist at the Council for Scientific and Industrial Research, the largest South African research and development organization: "This is not just a feel-good issue. Very soon we will be in a world which will punish carbon economies". Domestic mitigation strategies, the article concludes, should rely on the country's ability to generate electricity with minimal burning of fossil fuel (coal); a goal considered achievable by 2050. However, these initiatives face obstacles including a lack of administrative power at local levels

science/2013/09/27/sa-needs-to-find-innovative-ways-to-reduce-cost-of-renewable--energy-sourceshttp://www.bdlive.co.za/national/science/2013/09/27/sa-needs-to-find-innovative-ways-to-reduce-cost-of-renewable-energy-sources

[10] Gosling, M. (2014). Swift action vital to avoid disaster. *Cape Times*, April 2. Retrieved from   http://www.iol.co.za/scitech/science/environment/swift-action-vital-to-avoid-disaster-1.1669813#.Vdw2ulmDTjNhttp://www.iol.co.za/scitech/science/environment/swift-action-vital-to-avoid-disaster-1.1669813#.Vdw2ulmDTjN

and insufficient funding for such projects. The voice of Debra Roberts, a municipal environmental planner, is used to reinforce the discourse of *policy efficiency* through references to changes in the ways cities are designed and governed.

CONCLUSION

The analysis of country differences and similarities follows the BRICS rationale and media systems approach presented at the beginning of the chapter. The findings provide a better understanding of how the use of media space helps to formulate and justify climate policies.

First, a commonality among the BRICS countries is the explicit primacy of national economic discourses ranging from attempts to demonstrate how industries can help address climate change to more general suggestions on how to develop "green technologies" and cleaner energy resources (biofuel, natural gas). In Brazil and Russia, the calls for systemic change, albeit not strongly articulated, exemplify a trend toward reconsidering fundamental economic principles that have had negative impacts on the climate. The studied materials also imply that in public discourse there are moments when scientific arguments criticize power-related issues of economy and energy. The solutions in the examples from China and South Africa do not suggest such radical changes and instead call for pragmatic adjustments to the system, complying with the existing order rather than challenging it.

Second, although the media's use of scientific voices is dominant, NGO spokespersons sometimes play the role of "proxies" for science, changing the status of the scientific evidence (by taking a more subjective angle) and helping to better explain what is at stake. However, the role of the media (whether it is critical, an "independent" watchdog or a facilitator of governmental policies), as well as the place that the chosen newspapers occupy in the national media systems, provides a useful explanatory framework for future research. Based on this chapter, questions regarding the reasons why particular voices express certain positions and what this means for policy making processes are worth studying further.

Third, examples from the IPCC coverage reveal that there are many faces of the BRICS as an alliance of emerging economies. Depending on the basis for comparison, the members of the group can appear to have much in common in some cases and look radically different in others. Some of the observations indicate similarities that allow us to discuss these countries in pairs rather than as a unified, homogeneous block. For example, the pairing of Brazil and South Africa differs from the pairing of Russia and China. The former is more open to domestic political critique and calls for policy changes. In the latter, coverage is less critical, mostly highlighting national strategic interests with less concern for particular local problems. This represents an important dividing line for thinking about what kind of journalistic professionalism prevails in these countries, and what communicative cultures are being cultivated by the BRICS. In this sense, presently the BRICS concept fails to capture a strong group identity.

There was one thing missing in media coverage of the IPCC report in each of the studied countries. When media were concerned with scientific evidence that underlines the universality of climatic predictions and possible solutions, we did not see references to common goals or cooperation among the BRICS group. This contrasts with what happened at the meeting of BRICS environment ministers in Moscow. There, as mentioned in the introduction, delegates raised particular issues regarding technology exchange and economic efficiency. Pursuing one's own concrete interests through inter-group mutuality resembles the formation of country blocks during the intergovernmental climate negotiations (COP). However, for a global discourse of science, as we observed in this chapter, practices of coalition building are less common (even among established geopolitical groups such as BRICS). It is more relevant at the stage when science turns into actual policy.

Bibliography

Ackerman, F., & Stanton, E. A. (2008). The cost of climate change: What we'll pay if global warming continues unchecked. Released by Natural Resources Defense Council. Retrieved from http://www.nrdc.org/globalwarming/cost/cost.pdf

Andonova, L., & Alexieva, A. (2012). Continuity and change in Russia's climate negotiations position and strategy. *Climate Policy, 12*(5), 614–629.

Becker, U. (2013). *The BRICs and emerging economies in comparative perspective: Political economy, liberalisation and institutional change.* New York: Routledge.

BRICS. (2015, April 22). *BRICS to establish a working group on environmental protection.* Official website of Russia's presidency in BRICS. Retrieved from http://en.brics2015.ru/news/20150422/62265.html

Clark, D. (2011). Which nations are most responsible for climate change? *The Guardian*, 21 April. Retrieved from http://www.theguardian.com/environment/2011/apr/21/countries-responsible-climate-change?

de Albuquerque, A. (2012). On models and margins: Comparative media models viewed from a Brazilian perspective. In D. C. Hallin & P. Mancini (Eds.),

Comparing media systems beyond the Western World (pp. 72–95). New York: Cambridge University Press.

de Coning, C., Mandrup, T., & Odgaard, L. (2014). *The BRICS and coexistence: An alternative vision of world order.* New York: Routledge.

d'Essen, C. (2010). Brazil: COP15 as a political platform. In E. Eide, R. Kunelius, & V. Kumpu (Eds.), *Global climate, local journalisms: A transnational study of how media make sense of climate summits* (pp. 83–96). Bochum: Projekt Verlag.

EDGAR. (2013). *CO2 time series 1990–2013 per capita for world countries.* EDGAR Emission Database for Global Atmospheric Research, European Commission. Retrieved from http://edgar.jrc.ec.europa.eu/overview.php?v=CO2ts_pc1990-2013

Germanwatch. (2015). Global climate risk index 2015. Released by Germanwatch.org. Retrieved from http://germanwatch.org/en/download/10333.pdf

Hadland, A. (2012). Africanizing three models of media and politics: The South African experience. In D. C. Hallin & P. Mancini (Eds.), *Comparing media systems beyond the Western World* (pp. 96–118). New York: Cambridge University Press.

Hallin, D. C., & Mancini, P. (2004). *Comparing media systems: Three models of media and politics.* New York: Cambridge University Press.

Hallin, D. C., & Mancini, P. (2012). *Comparing media systems beyond the Western World.* New York: Cambridge University Press.

International Monetary Fund. (2013). World economic outlook database. Released by IMF. Retrieved from http://www.imf.org/external/pubs/ft/weo/2015/01/weodata/index.aspx

ITU. (2013). Global ICT developments. Released by International Telecommunication Union. Retrieved from http://www.itu.int/en/ITU-D/Statistics/Pages/stat/default.aspx

Jakubowicz, K. (2007). *Rude awakening: Social and media change in Central and Eastern Europe.* Cresskill, NJ: Hampton Press.

Leal-Arcas, R. (2013). The BRICS and climate change. *International Affairs Forum, 4*(1), 22–26.

Lee, T. M., Markowitz, E. M., Howe, P. D., Ko, C.-Y., & Leiserowitz, A. A. (2015). Predictors of public climate change awareness and risk perception around the world. *Nature Climate Change, 5,* 1014–1020.

Liu, S. (2011). Structuration of information control in China. *Cultural Sociology, 5*(3), 323–339.

ND-GAIN. (2013). The Notre Dame global adaptation index. Retrieved from http://index.gain.org/matrix

Nordenstreng, K., & Thussu, D. K. (2015). *Mapping BRICS media.* New York: Routledge.

O'Neill, J. (2001). Building better global economic BRICs. *Global Economics Paper No. 66,* released by Goldman Sachs. Retrieved from http://www.goldmansachs.com/our-thinking/archive/archive-pdfs/build-better-brics.pdf

Orgeret, K. S., & d'Essen, C. (2012). From COP15 to COP17. Popular versus quality newspapers: Comparing Brazil and South Africa. A question of social responsibility? In E. Eide & R. Kunelius (Eds.), *Media meets climate: The global challenge for journalism* (pp. 263–280). Göteborg, Sweden: Nordicom.

Poberezhskaya, M. (2015). Media coverage of climate change in Russia: Governmental bias and climate silence. *Public Understanding of Science, 24*(1), 96–111.

Reporters Without Borders. (2015). *World press freedom index*. Released by Reporters Without Borders. Retrieved from http://index.rsf.org/#!/index-details

Roudakova, N. (2008). Media-political clientelism: Lessons from anthropology. *Media, Culture & Society, 30*(1), 41–59.

Rowe, E. W. (2012). International science, domestic politics: Russian reception of international climate-change assessments. *Environment & Planning D: Society & Space, 30*(4), 711–726.

Saleh, I. (2012). Ups and downs from Cape to Cairo: The journalistic practice of climate change in Africa. In E. Eide & R. Kunelius (Eds.), *Media meets climate: The global challenge for journalism* (pp. 49–65). Göteborg: Nordicom.

Schreiner, W., & Bosman, J. (2012). Coverage cop-out: Global media analysis points to a lack of climate change coverage. *Ecquid Novi: African Journalism Studies, 33*(1), 66–71.

Sparks, C. (2014). Deconstructing the BRICS. *International Journal of Communication, 8*, 392–418.

Statistics Norway. (2013). Statistisk sentralbyrå. Retrieved from http://www.ssb.no/en/

Straubhaar, J. (2015). BRICS as emerging cultural and media powers. In K. Nordenstreng & D. K. Thussu (Eds.), *Mapping BRICS media* (pp. 87–103). New York: Routledge.

UN Development Programme. (2013). *Human development reports by United Nations Development Programme*. Retrieved from https://assignmentbro.com/blog/human-development-index-and-its-components

Vartanova, E. (2012). The Russian media model in the context of Post-Soviet dynamics. In D. C. Hallin & P. Mancini (Eds.), *Comparing media systems beyond the Western World* (pp. 119–142). New York: Cambridge University Press.

Wu, F. (2012). Sino–Indian climate cooperation: Implications for the international climate change regime. *Journal of Contemporary China, 21*(77), 827–843.

Xu, P. (2010). China: Emerging player with a historical legacy. In E. Eide, R. Kunelius, & V. Kumpu (Eds.), *Global climate, local journalisms: A transnational study of how media make sense of climate summits* (pp. 131–145). Bochum: Projekt Verlag.

Zhao, Y. (2012). Understanding China's media system in a world historical context. In D. C. Hallin & P. Mancini (Eds.), *Comparing media systems beyond the Western World* (pp. 143–173). New York: Cambridge University Press.

CHAPTER 8

Who Captures the Voice of the Climate? Policy Networks and the Political Role of Media in Australia, France and Japan

*Shinichiro Asayama, Johan Lidberg, Armèle Cloteau,
Jean-Baptiste Comby, and Philip Chubb*

CLIMATE CHANGE, "WICKED PROBLEM" AND THE CONCEPT OF NETWORKS

The complexity and gravity of climate change have compelled some researchers to label it a "wicked problem" (Rayner 2006). There is no single or clear definition of the limits of the problem and hence no ultimate single solution to climate change. Climate change is a conflict between the market

S. Asayama (✉)
National Institute for Environmental Studies, Tsukuba, Japan

J. Lidberg • P. Chubb
School of Media, Film and Journalism, Monash University, Melbourne, Australia

A. Cloteau
Department of Sociology and Political Science, Paris-Saclay University,
Paris, France

J.-B. Comby
French Press Institute, University of Paris, Paris, France

© The Author(s) 2017   171
R. Kunelius et al. (eds.), *Media and Global Climate Knowledge*,
DOI 10.1057/978-1-137-52321-1_8

and the environment, between rich and poor countries and between present and future generations, just to mention some of the obvious dimensions (Chap. 1). Because such fundamental stakes are at play, and due to the different ideas, values and worldviews involved in the debate (Hulme 2009), it is immensely difficult to reach a consensus on specific policy responses to climate change. A particular factor in this complexity is the fact that agreement on global targets for greenhouse gas (GHG) emissions reduction and on binding policy decisions—either in international treaties or in regional governance mechanisms—must be bargained and determined by local political institutions (mostly at the national level). This fact of "post-national" twenty-first century politics highlights the importance of situating media coverage into the framework of national political systems.

From a policy perspective, governments do not have a clear hegemonic power to steer the policy-making, and the "wickedness" of climate change as a problem further complicates political decision-making. In dealing with climate policy-making, governments are dependent on complex networks of various actors with specific interests and resources such as industries, environmental NGOs, scientists/experts in the climate field and so forth. Here, the concept of "policy networks" becomes valuable for the analysis of climate policy. It enables analysts to "…captur[e] the dynamic and complex relations involved in the climate change policy process" (Bulkeley 2000). Such analysis starts from the assumption that policy formation crucially takes place in the *informal* networks between governments and other interested actors, and that policy outcomes are determined by negotiations and resource exchanges within such networks (Blanco et al. 2011; Bulkeley 2000; Compston 2009). The notion of network emerged as part of the broader theoretical shift "from government to governance," because it resonates with the real world political decisions made in the process of interaction between stakeholders. This, in turn, attracted attention in the policy studies literature (Chilvers and Evans 2009). This network concept is also employed to articulate the emergence of a new governance paradigm, "network governance," and it is often confused with "policy network" theory (Blanco et al. 2011; Chilvers and Evans 2009). We are, however, interested in describing the local political fields where actual policy-making takes place and where the media play a certain political role. We use the notion of network within the "policy network" framework.

Policy networks operate through the exchange of resources between members of a network. Compton (2009) argues that the crucial role of media within policy networks is "…to change the policy preferences of other actors by altering their perceptions of problems, solutions or other aspects of the policy process." The media is an actor in the network. In this capacity it has at least two functions. It enables others in the network to make their claims. Further, as Olausson and Berglez (2014a) note, media act independently to "construct and negotiate meanings." This process can be highly competitive but also routinized. For particular members of the network, media exposure can be both an advantage and a disadvantage. Understanding who enters the media to speak about climate assists in building knowledge of how media operate in climate policy networks (Olausson and Berglez 2014a). In this chapter, we compare voice representation in the IPCC AR5 coverage across three countries. These are Australia, France and Japan. We contextualize this media analysis into national political fields by sketching out local policy networks. Such comparative analysis provides fertile ground for thinking about distinct kinds of policy networks and the media's role in each country. As we shall see below, public discourses on climate change have taken very different paths in Australia, France and Japan.

Analyzing media coverage does not, of course, offer direct evidence about policy networks. In fact, the close empirical analysis of policy networks goes beyond our scope. This chapter does not aim to draw a holistic picture of climate policy networks in each country, nor create a new theory of policy networks. Instead, we use the concept broadly to build a perspective from which to make sense of media coverage in relation to policy-making. Our description of policy networks is a basic sketch of national political fields and largely based on our knowledge of local political and media contexts.[1] Nevertheless, this offers a chance to develop a more detailed discussion about the role of media in the national political fields or networks in which decision-making takes place. In order to better understand the nuances and dynamics of the "domestication" of global climate change (see also Chap. 4)—translation of the IPCC reports or climate science to national political networks—we examine findings from the media analysis in the context of these locally specific networks (and political cultures). This helps to understand how news coverage is constructed in local political contexts. More importantly, it helps to address questions about the media's role in the complex nexus of science–policy–media networks in different countries.

[1] It is important to note here that we all have different epistemological and disciplinary backgrounds, thus the analysis of our respective national contexts is framed in slightly different ways: Comby/Cloteau (field theory/Bourdieu), Asayama (science & technology studies) and Lidberg/Chubb (journalism and policy studies). While these are different perspectives with somewhat different theoretical underpinnings (in relation to the conscious-rationality of actors, for instance), what we have in common is a focus on the complexity of networked relations that form the actual infrastructure of power and bargaining in political decision-making.

Climate-Energy Politics and Journalistic Cultures

Here, we first briefly draw out the local backgrounds of energy and climate politics, combined with different journalistic cultures and media systems in three countries. This contextual background provides necessary information for the following discussion on the local policy networks and the role of media within them. Table 8.1 provides a brief summary of each country's context.

Table 8.1 Overview of political and media context on climate change

Country	Political and media context on CC	
	Climate and energy politics	*Journalistic culture/media system*
Australia	*Climate* • Contentious politics on legislation and repeal of the carbon price • Climate skepticism in the Liberal conservative government *Energy* • Heavy reliance on coal as domestic energy	*Journalistic culture* • "Liberal model"—strong commercial tradition and information-oriented journalism combined with a strong public broadcaster *CC in the media* • Largest media owner (Murdoch's News Corp Australia) actively discrediting climate science and campaigning against climate actions
France	*Climate* • CC is a consensual issue—high profile of climate scientists • CC is at high political agendas—urgent demand but little political initiatives *Energy* • Historical hegemony of nuclear power since 1970s	*Journalistic culture* • "Pluralist model"—political parallelism and commentary-oriented journalism *CC in the media* • No climate skepticism in the media • Specialized "climate journalists"—main focus on science, little coverage on economy/policy
Japan	*Climate* • Bipartisan consensus on the reality of CC • A prolonged inter-ministerial battle over climate change legislation *Energy* • Polarized views on the role of nuclear power (Post-Fukushima phase)	*Journalistic culture* • "Press club/corporatist model"—segmented pluralism with beat journalism and pack journalism through the press club system *CC in the media* • No climate skepticism in the media • Ideological divide on mitigation policy and nuclear power

Australia: Contentious Politics, Vocal Climate Skeptics and Heavy Reliance on Coal

In July 2014—when the AR5 Synthesis report was published—the Australian Parliament repealed 2011 legislation establishing a carbon pricing policy and promoting investment in renewable energy. This globally unique turn-around of climate policy occurred because of Australia's heavy reliance on coal, one of the largest foreign exchange earners in the country (Hannam 2014). The repeal of carbon pricing can be seen as the culmination of one of the most divisive periods in Australian political history as key actors in the climate policy network divided into warring tribes (Chubb 2014). The carbon price was initially proposed by the Liberal-National Party government under Prime Minister John Howard, for whom it was a pledge during the election of November 2007. Howard was defeated by Labor leader Kevin Rudd in that election, who also favored carbon pricing. The initiative thus had overwhelming bipartisan support, with even the business sector willing to compromise in the face of high public climate concern. In October 2011, new Labor Prime Minister Julia Gillard succeeded in passing the carbon pricing bill. However, the bipartisan support for the carbon price had by then broken down. This occurred in December 2010 when the Liberal-National Party elected a new leader, Tony Abbott, who came to power at the election in September 2013. Abbott and many members of his Liberal Party were skeptical of the reality of anthropogenic climate change. They argued that Australia should rely on coal, both for cheap domestic energy and for export dollars. This led to a decision to revoke the carbon price, claiming it had been based on anti-growth "green" ideology (Taylor 2011).

The state of the Australian media system partly helps to explain the political divisions on climate policy. During 2007–2008, when there was still bipartisan support for the carbon price, major newspapers—including the *Sydney Morning Herald* and the *Daily Telegraph*—echoed the public enthusiasm for climate action. However, as Kevin Rudd turned away from implementing the carbon price in 2010, there was a marked shift in editorial tone and content of news coverage (Manne 2012; Chubb and Bacon 2010; Chubb 2012, 2014). The media played a major role in undermining Rudd's leadership by focusing on the decline in his popularity, fanning the discontent that genuinely intensified confrontation within the Labor government. There is an extreme concentration of media ownership in Australia, with Rupert Murdoch's News Corp controlling more than half the newspapers and other media outlets. These outlets helped cast doubt on the reality of anthropogenic climate change (Flew 2013; McKnight 2012).

France: Consensus Science, Market-Based Political Actions and the Established Hegemony of Nuclear

In France, climate change has been viewed as a global issue best dealt with by using market-based policy. Public awareness of climate change increased in the 2000s. Climate change gained a higher political profile, climbing to the top of the legislative agenda, after the "Grenelle Environment" was launched by former President Nicolas Sarkozy in 2007, bringing different actors (government, industry, civil society groups, etc.) to the table to discuss key issues regarding the environment. Consensus grew strong on the need for action to counteract anthropogenic climate change and there has been little room left for climate skepticism. This consensus framing of climate science stems partly from the historical background of climate science in France (Aykut and Dahan 2015). Climate scientists themselves progressively gained a robust legitimacy in French society and their field of work was institutionalized in the late 1990s. However, French climate policy has made slow progress with only a few policies, such as tax credits and public awareness campaigns, implemented thus far. In particular, France's high dependency on nuclear power hampered policy incentives for expanding renewable energy. Noticeably nuclear power has enjoyed priority in French energy politics, since political choices were made in the 1970s. There is, thus, a long and strong historical and structural political hegemony and strategic consensus around nuclear power, which has also defined nuclear power as a clean and safe choice (Lascoumes 1994; Lacroix and Zaccaï 2010). In terms of concrete action, then, little has been achieved in climate policy in France. There is disagreement on what to do and how. This is fueled by increasing criticism, from some political left-wing organizations and the Green Party, on the role of nuclear power in the future energy mix. To balance this point, the politics of climate change in France is characterized by increasing concern and debate, but limited and slow policies to address the issues.

French media representation has largely reflected the political trajectory of climate policy in France. Media enhance the consensual view on human-caused climate change and skeptical voices are virtually nonexistent in the French media (Comby 2015). Climate news coverage is densely professionalized. A handful of "climate journalists" who specialize in environmental issues write the vast majority of the stories. These climate journalists are generally inclined to quote natural scientists as their main sources, while economic or policy issues on climate change are relatively untouched. The French government's campaigns to raise public awareness and push for solutions based on individual behavioral change are uncritically portrayed. As a consequence of all this, the lack of political initiatives on climate action is generally left unquestioned.

Japan: Authorized Science, Political Battle on Mitigation Targets and Polarized Nuclear Debate

Climate politics in Japan has been largely characterized by the prolonged battle between the environment and economic ministries over controlling climate policy-making (Schreurs 2002), as well as by the symbolic power of "Kyoto" (Tiberghien and Schreurs 2007). The very fact that the name of the Japanese ancient capital was used to label the Kyoto Protocol (KP) put the normative load on taking climate actions. Two critical events have recently influenced climate politics in Japan. First, after the 2009 election, the Democratic Party of Japan (DPJ) led government proclaimed an ambitious target for GHG emissions reduction (25 percent by 2020 from 1990 levels), coupled with the introduction of carbon tax and cap-and-trade schemes. The proposed target caused extensive public controversy. It met strong opposition from industry and even from trade unions, the DPJ's largest political supporter. Adding to the complexity, the target was largely based on the expansion of nuclear power. However, the Fukushima nuclear disaster in 2011, the second critical event, radically changed the course of Japanese climate policy. The disaster necessitated the fundamental review of pre-Fukushima policy, forcing the DPJ government to conduct, for the first time, a national debate on energy and climate issues through several public forums. "Deliberative polling" methods were also used (Skea et al. 2013). Against a backdrop of strong public support for a nuclear phase-out, the DPJ government proposed a new climate and energy policy, laying out the goal to phase out nuclear power by the end of the 2030s (Skea et al. 2013; Kuramochi 2015). However, after a landslide victory in the general election of December 2012, the new conservative government, led by the Liberal Democratic Party (LDP), abandoned this policy, shifting back to a course more aligned with pre-Fukushima policy. It emphasized the role of nuclear power and called for the continuation of existing plants, and even implied the building of new nuclear power plants (Kuramochi 2015). At the same time, the new conservative government revised and toned-down the mitigation target, allowing an increase in GHG emissions to 3.1 percent by 2020 from 1990 levels (Kuramochi 2015). Until now, the current LDP government has been keen to keep nuclear power as a central source of energy supply in Japan, while showing less interest in taking further climate change policy actions.

The Japanese media's portrayal of climate science tends to be quite uniform among different newspapers. The IPCC has been framed as the authorized scientific voice to speak for the climate, and no significant media space has been given to climate skeptics (Asayama and Ishii 2014). However, in reporting mitigation policy, particularly the role of nuclear power, the media are politically divided. Left-leaning newspapers were critical of nuclear power, backed the KP and advocated mandatory measures (e.g. cap-and-trade). Conservative papers were critical of the efficacy of the Kyoto treaty, favoring voluntary actions by industry and strongly advocating for nuclear power as a way to mitigate climate change.

Journalistic Cultures in Australia, France and Japan: Liberal, Pluralist or Corporatist Model

Australia, France and Japan have three distinct journalistic cultures or traditions that structure and determine the relationships between the media and political systems. These journalistic cultures are important as a means to understand the link between policy networks and the role of media in each country's local context. The seminal work by Hallin and Mancini (2004) on three ideal models of media and political systems can be illustrative here to explain the journalistic cultures in our three cases. Australia can be classified as a *liberal model*, mainly due to its historical association with the USA and Britain. The liberal system can be characterized by a predominance of commercial or market-driven media organizations and information-oriented journalism, although the commercial aspect of the Australian media model is somewhat balanced by a strong and well-funded national public broadcaster (the Australian Broadcasting Corporation).

French journalistic culture bears more resemblance to the *pluralist model*, where the role of the state (public subsidies) is strong, political parallelism (between media and politics) and a strong tradition of commentary-oriented (or advocacy) journalism prevail. The French media field has been structured by a right-left political division between *Le Figaro* and *Le Monde* on one side and *Libération* and *L'Humanité* on the other. Although this division has become less severe in the last ten years, the right–left opposition is still a strong feature of French journalistic culture.

While the media-politics dynamic in Japan has many particular features, it can be seen as a specific variation of the *corporatist model*. Japanese journalism carries a tradition of strong "segmented pluralism" which is structurally organized as rigid beat journalism. Furthermore, the Japanese journalistic culture is notably characterized by the infamous press club (*kisha club*) system, an association of major newspaper reporters assigned to key social and political organizations such as the government agencies. Membership provides convenient access to official information sources. The *kisha club* system nurtures a cozy and close relationship with political and economic elites, institutionalizing a culture of "pack journalism" (Feldman 2011). Thus, Japanese journalistic culture can be characterized as a *press club/corporatist model.*

It is, however, worth noticing that even though genuine variations exist in the journalistic cultures of the countries under study, in many countries there is a growing tendency to move toward the *liberal model.* This is mainly due to the rapid increase in media commercialization and market-driven journalistic culture.

VOICE REPRESENTATION IN THE IPCC AR5 COVERAGE

One crucial role of media in the policy process is the "gatekeeping" function of determining who gets access to the media, speaking for the issues at stake. As Compton (2009) said, the media can ideally change the condition or the rule of game where actual policy processes take place. The political power of the media is notably exerted when defining *who* is an authorized voice. By determining who can speak for the climate, the media can *informally* influence formal climate policy-making, either by contributing to government policy or, at times, by obstructing it, depending on the political position of the media in complex climate policy networks. In this chapter, we look more closely at the voice representation in AR5 coverage in each country.[2]

Table 8.2   Number of relevant stories/voices and newspapers' political leaning

Country	Stories and voices			Newspapers	
	All stories	Relevant stories	Relevant voices	Liberal or left-leaning	Center-right or conservative
Australia	275	65	157	Sydney Morning Herald	Daily Telegraph
France	85	52	76	L'Humanite	Le Monde
Japan	171	63	73	Asahi Shimbun	Yomiuri Shimbun

"All stories" = the number of news articles in total sample

"Relevant stories" = the number of news articles in high relevance to the IPCC AR5 (either "IPCC AR5 report is the main story" or "general climate story with reference to IPCC AR5")

"Relevant voices" = the number of voices quoted in "relevant stories"

[2]We must acknowledge our debt to Dmitry Yagodin for his assistance analyzing "voice representation" in each country.

Table 8.2 shows the number of news articles and voices represented in the newspapers under study. Our total sample of AR5 news coverage includes not only stories in which the AR5 is a main topic, but also general climate stories with no references to the AR5 and/or stories that simply mention climate change or the IPCC (Chap. 1). However, since the focus of our analysis here is exploring the link between the media and policy networks around the AR5, we confined our sample to "relevant stories"; that is, news articles with the AR5 report as the main story or general climate stories that refer to the AR5.

The two newspapers chosen in each country have different political orientations, covering a broad spectrum of ideological positions pertaining to climate change—from *neoliberal* or *conservative* stances, putting the economy ahead of the environment, to *liberal/progressive* positions emphasizing the urgency of climate actions.

To identify who gets to speak in the media, we compared the results of "voices" coding. "Voices" are defined as *people* who are quoted either directly or indirectly. Importantly, they are only coded when their names are included. All coded voices are then classified into different categories based on how their respective roles and positions were referred to in the articles (for further detail on coding "voices" see Chaps. 1 and 3). We analyzed two aspects of voice representation: *dominance* (who was given the power to speak) and *diversity* (the variety of voices represented). The summary of the results is shown in Tables 8.3–8.5.[3]

Table 8.3   Distributions of "voice categories" from relevant stories

Country	Voice category (%)						Total (%)
	National politics	Transnational politics	Civil society	Business	Science	Media	
Australia (n = 157)	22%	15%	13%	4%	34%	12%	100%
France (n = 76)	13%	5%	1%	7%	74%	0%	100%
Japan (n = 73)	14%	7%	10%	3%	64%	3%	100%

n = the number of "relevant voices" (the voices quoted in the news articles in high relevance to the IPCC AR5)

[3] In Table 8.5 there is a relatively high percentage of "no voice" stories both in France and Japan, compared to Australia. This may reflect the different journalistic practices in France and Japan, where it is more common to use unnamed news sources in reporting on political events.

Table 8.4 Distribution of domestic/foreign voices from relevant stories (percent)

Voice category		Australia (n = 157)	France (n = 76)	Japan (n = 73)
Domestic voices		54	66	37
	Science	26	49	25
	Politics	15	10	1
	Business	3	7	3
	Civil society	10	0	8
Foreign voices		19	27	53
	Science	8	23	40
	Politics	7	3	12
	Business	1	0	0
	Civil society	3	1	1
Transnational politics		15	6	7
Media		12	0	3

n = the number of "relevant voices" (the voices quoted in the "relevant stories")

Table 8.5 Percentage of relevant stories with specific "voice combinations" (percent)

Combination of voices	Australia (n = 65)	France (n = 52)	Japan (n = 63)
Science	18	46	29
Science + media	5	0	0
Science + civil society	0	0	2
Politics	18	10	5
Politics + media	5	0	0
Politics + civil society	5	0	2
Science + politics (+ others)	25	10	11
Media	3	0	0
Business	2	2	0
Civil society	17	0	5
No voice	3	33	48
Total	100	100	100

n = the number of "relevant stories" (the news articles in high relevance to the IPCC/AR5)

"Politics" includes the voices of both "national politics" and "transnational politics"

"Other" includes the voices of either "media", "business" or "civil society"

Dominance of the Voices: Whose Voices were Authorized and Dominant?

As seen in Table 8.3, in the French and Japanese media the dominant voice is *scientists*. The large majority of relevant voices came from the scientific domain—over seven-out-of-ten and nearly two-thirds of the total voices were in the "science" category in France and Japan, respectively. Interestingly, while French and Japanese voice representation resembled each other in the dominance of scientific voices, they were quite different in terms of a domestic–foreign comparison. Table 8.4 shows that in France, domestic voices, particularly *domestic scientists* such as Hervé Le Treut, a French climatologist taking a role of review editor in the AR5 WGII report, had a hegemonic position. In contrast, foreign voices dominated over domestic ones in Japan. In fact, *foreign scientists* such as the then IPCC chairman, Rajendra Pachauri appeared most frequently in the Japanese media. In the case of Australia, the dominance of science was relatively mild, and there was no strong hegemonic voice compared to France and Japan. Rather, *political actors* were more prominent than scientific ones in Australia—as seen in Table 8.3, nearly 40 percent of the total voices were either in the "national politics" or "transnational politics" categories, followed by "science." Also, more than a half of the total stories referred to political voices in Australia whereas those stories were less prominent in France and Japan (20 percent and about 30 percent respectively, see Table 8.5). In particular, a *domestic politician*, federal environment minister Greg Hunt, was the most frequently cited, illustrating the relatively strong dominance of domestic political sources in Australian coverage of the AR5.

Diversity of the Voices: How Diversified is the Voice Representation?

As noted, the representation of voices in France and Japan was overwhelmingly concentrated on scientists (Table 8.3). This result is also clear in Table 8.5, which shows the percentage of stories with different combinations of the voices represented concurrently within a single news article. In France and Japan, when quoted in the media, scientists most often appeared independently, not together with other voices. This tendency is particularly noticeable if we look at the stories with only domestic voices—of the total stories mentioning domestic voices (23 and 16 articles in France and Japan respectively), the large majority of stories (more than

70 percent in France and 60 percent in Japan) only contained scientific voices. As such, the French and Japanese media had a low diversity of voices. Interestingly the Australian media had a far more diverse representation of voices. The voices from "civil society" and "media" categories had a higher presence in Australia, compared to France and Japan (Table 8.3).[4] Overall Australia had greater voice diversity compared to France and Japan, and most notably, stories referring to both scientific and political voices had the largest share of the total stories in Australia (Table 8.5). That is, scientists most commonly appeared in the Australian news if coupled with political voices. This result perhaps indicates the contentious or competitive nature of voice representation in Australia.

Overall, scientific voices were overwhelmingly dominant in France and Japan. This led to low voice diversity, while in Australia political voices had relatively high prominence, resulting in more diversity or competition in voice representation. This can be interpreted to suggest that climate science is firmly settled and authorized in France and Japan, where the climate debate is captured strongly by the scientific perspective. On the contrary, the scientific basis of climate change is argued over and politicized in Australia, thus highlighting political pluralism in the climate debate.

POLICY NETWORKS AND THE ROLE OF MEDIA IN THE NETWORKS

According to Bulkeley (2000), types of policy networks vary, depending on the nature and pattern of the interaction between actors involved. There are two main variations of policy networks: *policy community* and *issue networks*. *Policy community* is a strongly institutionalized and closed type of network in which there are shared values, worldviews and policy goals and where a limited number of members agree on a specific policy response to the problem in question. On the other hand, the *issue network* is less institutionalized and more open, typically harboring a broader membership relevant to policy-making, and thereby producing less agreement on the issue at hand and the way it should be addressed.

Based on our knowledge about each country's political and media contexts pertaining to climate change, we describe what types of policy network exist in each country. Then, we define the role of media in the networks

[4] It is worth noting that "business" voices were commonly limited in all three countries – this is perhaps because relevant stories in the IPCC AR5 coverage focused more on science and climate change impacts rather than its economics, hence leaving less opportunity for business actors to speak.

based on the analysis of voice representation in the IPCC AR5 coverage. As noted above, the description of policy networks here is not based on empirically grounded analysis. Our aim is rather to understand the political functions of media in local political contexts. With this in mind, we use the concept of policy network as an analytical lens to illustrate local political contexts and contextualize the media analysis. Table 8.6 shows the overview of policy networks and the media's role in three countries.

Australia: Two Opposing Networks and Media Polarization on Carbon Price

In Australia, we can identify at least two opposing policy networks: the *"fossil fuel lobby coalition"* (policy community) and the *"sustainable economy network"* (issue network). The "wicked" nature of climate change is exacerbated by the nation's reliance on coal. Eighty percent of national electricity is obtained from coal-fired power stations. Australia is one of the biggest GHG emitters per capita among developed nations. At the same time, coal is one of the nation's biggest export earners (Chubb 2014). The *fossil fuel lobby coalition* is built upon strong vested interests in maintaining fossil fuel dependency. It therefore makes every effort to block climate policies such as carbon pricing (Bailey et al. 2015). Because the members of this coalition are narrowly confined to the fossil fuel industry and they share common interests and policy goals, the nature of this coalition is the strongly institutionalized *policy community.*

On the other hand, the *sustainable economy network* consists of a number of NGOs and independent climate action advocacy groups, along with the Green Party. A significant part of the corporate sector in Australia has supported a carbon price and displayed a willingness to transform from a fossil fuel-based economy into a new sustainable and renewable economy. However, to implement such complex policy, broad-based multi-party backing is needed. Thus, the nature of this network is a diverse and fluid *issue network*—there was only a short-lived multi-party committee set up under former Prime Minister Julia Gillard. The polarization of debate within this network is a manifestation of policy breakdown and, thus far, a complete failure of consistent and communicative political leadership on climate change.

Previous studies (Chubb 2012; Chubb and Bacon 2010; Bacon 2011) have several times verified how the conservative Australian media (most notably *News Corp*, the owner of the *Daily Telegraph*) clearly misrepresented the "scientific consensus" on anthropogenic climate change by

Table 8.6 Policy networks and the role of media in the networks

Country	Policy networks			Media's role in networks
	Name	Type	Shared concerns or policy goals	
Australia	"Fossil fuel lobby coalition"	Policy community	• Keep fossil fuel dependency (especially, coal) • Block any climate actions to regulate economic growth (e.g. carbon price)	• Undermine "scientific consensus" of ACC • Cause political failure of legislating carbon price *Polarization of climate policy*
	"Sustainable economy network"	Issue network	• Transformation into the sustainable renewable energy economy • Support for carbon pricing policy	
France	"Green growth expert circle"	Policy community	• CC is a political priority and a global issue • Nuclear is a main short term solution in France • "Green growth"—the environment as an opportunity to boost economy	• Climate scientists connect the media and "expert circle" • Specialized "climate journalists"—legitimize climate science • Give less space to critical/ alternative voices from "ecology network" *Reproduction of elite science discourses*
	"Political ecology network"	Issue network	• CC is a political priority • Engage in alternative social movement—climate actions should be at societal level	
Japan	"Nuclear lobby coalition"	Policy community	• Nuclear is harmful, not clean energy • Continuous use and expansion of nuclear power • CC is on political agenda but a secondary issue	• Create common understanding that CC is happening among the networks • *Yomiuri* advocates for "nuclear lobby" while *Asahi* abandons nuclear as low-carbon energy *Polarization of nuclear power*
	"Climate risk network"	Issue network	• Minimize the impacts of CC • Both mitigation and adaptation are necessary • No consensus on what type of energy is "desirable" in the future	

adopting what Boykoff and Boykoff (2004) call a "balance as bias" practice, leaving the public with an impression that there is still ongoing scientific debate on human-caused climate change. The conservative media's behavior has contributed to the polarization of climate policy in Australia. Since a complex policy, such as carbon pricing, demands wide agreement among the public on "humans being the cause of climate change" such framing not only undermines the public confidence in climate science but also causes political failure in attempts to legislate carbon pricing policy.

France: Two Continuum Networks and the Media's Reproduction of Elite Science Discourses

In France, there are two continuum policy networks, where the networks are organized as a continuum—hence less clearly demarcated. The *"political ecology network"* (issue network) is mainly composed of a large and historical social movement (Ollitrault 2008) and the Green Party. The *"green growth expert circle"* (policy community) is animated by engineers, climate scientists, think tanks and energy lobbies. The two networks are not necessarily in conflict and strong opposition, and membership in the networks commonly overlap. The distinction of these two networks is structured around and drawn by political choices on energy, particularly nuclear power. While the *green growth expert circle* emphasizes nuclear power as a main energy source for France, the *political ecology network* focuses on renewable energy as well as social mobilization. In the *green growth expert circle*, there is the underlying affinity between climate scientists and political/business elites, which creates a closed "sociability circle," sharing a global reformist view on the economy as well as a pro-nuclear approach. Regular cooperation among scientists, policymakers and economic stakeholders with shared worldviews characterizes this circle as a *policy community*. On the other hand, the *political ecology network*, characterized as a fluid and temporal *issue network*, portrays nuclear as a harmful energy source, and is sometimes engaged in radical contestation movements, such as protests against shale gas fracking, land settlements or urban planning. Despite the diverse issues at hand, this network commonly agrees that climate change is a political priority—a view shared with the *green growth expert circle*.

French climate scientists actively participate in the political mobilization to raise the social prominence of climate change (Aykut et al. 2012). This is crucial in understanding how the French media enter the policy networks. Climate scientists have a close relationship with a small group of journalists that specialize in environmental issues (e.g. Stéphane Foucart at *Le Monde* or Sylvestre Huet at *Libération*). They play an intermediate role by connecting these journalists to the *green growth expert circle*. The close link between scientists and specialized "climate journalists" overwhelmingly increases the media presence of scientists to claim climate urgency, while, at same time, leaving little room for dissonant voices from the *political ecology network*. In short, the French media functions politically to reproduce the elite scientific discourses, giving less space to critical and alternative voices. It is also worth noting that scientifically dominated news coverage causes policy issues to remain vague, providing a structural advantage to the *green growth expert circle* because it only marginally pressures the government into concrete climate change policies.

Japan: Two-Tiered Networks and the Media's Polarization of Nuclear Power

In Japan, there are two-tiered policy networks (Bulkeley 2000) where the former is partly accommodated by the latter: the *nuclear lobby coalition* (policy community) and *climate risk network* (issue network). The *nuclear lobby coalition* is a group with strong vested interests in nuclear energy, a so-called nuclear village (Kingston 2014). The nuclear village is a coalition of the power industry, nuclear researchers and conservative politicians, such as members of LDP with a specific interest in the spread of nuclear energy. Shortly after the Fukushima nuclear disaster in 2011, this *nuclear lobby coalition* lost political power, mainly because of strong public opposition to nuclear power. However, because the coalition is cohesive, well connected and shares strong common interests, it is resilient. It has retained the power to influence climate and energy politics (Kingston 2014; Vivoda and Graetz 2014), particularly after the conservative party, LDP, returned to office in 2012. As this coalition's primary focus is on nuclear energy, climate change is only a secondary issue used rhetorically to justify the continuation of nuclear energy.

The second policy network in Japan is the *climate risk network*, which is built on concerns about dire impacts of climate change such as sea level rise, floods, drought and general social unrest. Since the issue at stake is mostly future climate impacts, understanding the science of climate change becomes the foundation of this network and a common goal is to minimize the impacts of climate change. However, for this *issue network*, how to address climate policy is an open question. Which energy sources society should invest in are not agreed upon since this question involves diverse interests, values and political priorities on the part of different actors. As a common base line, the network appears to have recognized that due to the scale of the issue, mitigation and adaptation policies are needed to avert potential "catastrophic" climate consequences.

Both the *nuclear lobby coalition* and *climate risk networks* share a common understanding that anthropogenic climate change is happening and that it may have severe consequences for human society. The Japanese media enhance this common concern by authorizing climate science (in particular, the IPCC) and disseminating the results of the IPCC reports. However, in addressing policy responses, the two newspapers (*Asahi* and *Yomiuri*) played very different roles. While the *Asahi* delegitimizes and rejects nuclear power as a means of "low carbon energy," the *Yomiuri* justifies nuclear power as a necessary option for climate mitigation, thereby backing the *nuclear lobby coalition*. As a result, there exists polarization in how to tackle climate change, in particular over the future role of nuclear power.

MEDIATED DIVISION OF CLIMATE DEBATE AND THE NEED FOR NEW *BROKER*-JOURNALISM?

Australia, France and Japan provide interesting examples of local policy networks in action in the media and present a snapshot of the political fields of climate change politics. In all three countries there are at least two different policy networks built around energy-climate political decisions: (1) a *strategic policy community*, which tends to be built around a specific energy bloc (coal in Australia, nuclear in France and Japan) with high political investment and narrowly structured membership pushing a particular set of policy actions, and (2) an *alternative issue network* with heterogeneous members mostly organized around environmental risks and alternative economic pathways. However, the relationships between the networks vary in each country, depending on their local political histories and cultures. In Australia, the networks have an *oppositional* relationship— on one hand there is a strong coal industry hostile to any climate actions, denying scientific consensus, and on the other hand, civil society groups and some corporations seeking a new sustainable, renewable energy economy. The French networks are in a *continuous* relationship both in terms of prioritizing climate change and in terms of consensus on climate science. However, small communities of experts embrace a reformist view of green growth with pro-nuclear approaches whereas the large association of the political left centers on social mobilization through protest. In Japan the *tiered* structure of the networks is organized, the former as a coalition of interested industries and conservative politicians using climate rhetoric for nuclear advocacy. This, in turn, is partly accommodated by the latter, a broad network of actors bonded by concern over the risk of future climate impacts.

What then is the role of media in the local context in climate policy networks? Our analysis suggests that the media has widened the gap between existing policy networks, depending on each country's context, by giving different actors/stakeholders a voice to speak for the climate, especially when portraying climate science. In each country, the media seems to *disconnect* local policy networks from each other rather than bind them together. In Australia, the media politicized the debate on climate science by giving diverse voices a space to speak and creating a contentious atmosphere, which caused further ideological polarization. In France, the strong dominance of scientific voices in the media reaffirmed the prominence of the climate issue, but only marginally to pressure the government into taking concrete actions. French media also helped to reinforce the expert discourses enjoying political hegemony while, at the same time, marginalizing voices of dissent. Likewise, the high concentration of scientific voices in the media authorized climate science in Japan. However, what to do about climate change remained unclear, allowing flexibility in the use of science to mobilize opposing policy goals, both for nuclear advocacy and for a nuclear phase-out.

Taken together, in all three cases of Australia, France and Japan, despite differing degrees and dimensions, the media failed to connect the networks in local political systems and thereby intensified political polarization of the climate-energy debate. In other words, media are not yet able to facilitate and create the public discourses needed to negotiate the "wicked" policy pathways of global climate change. This, then, raises normative questions concerning *what* role the media *should* play under the "wicked" conditions of climate politics, and *how* the media can navigate politicized debate about climate change in different political settings. Reflecting the "wicked" nature of climate change, the immense complexity of definition and the inherent value clashes involved in decision-making, the media could be expected to search for a new mode of journalism that takes on an *intermediary* role, facilitating conversations among different stakeholders in national and global climate political fields. Nisbet and Fahy (2015) call this new journalism "society's new class of knowledge professionals" that can guide overly politicized science-related policy debate to bring together diverse yet conflicting views and defuse ideological polarization. They summarize its function using three "broker" roles, namely *knowledge broker, dialogue broker* and *policy broker*. We may call this new mode of journalism *broker*-journalism. As exemplified above, today's media and journalism operate inside their political comfort zone. This is a self-fulfilling ideological camp where they can meet the expectations of a particular audience who share an ideological standpoint and policy goal. However, such an ideologically divided media culture in itself has polarized and paralyzed dialogue on politically contentious issues such as climate change. To bridge this entrenched ideological divide and engage in critical conversation with dissident voices, there is a need for a new forward-looking *broker*-journalism that would help us understand why and what we disagree on and not alienate people but unite us in dealing with the climate crisis.

BIBLIOGRAPHY

Asayama, S., & Ishii, A. (2014). Reconstruction of the boundary between climate science and politics: The IPCC in the Japanese mass media, 1988–2007. *Public Understanding of Science, 23*(2), 189–203.

Aykut, S., Comby, J. B., & Guillemot, H. (2012). Climate change controversies in French mass media 1990–2010. *Journalism Studies, 13*(2), 157–174.

Aykut, S., & Dahan, A. (2015). *Gouverner le climat? 20 ans de négociations internationales.* Paris: Presses de Sciences Po.

Bacon, W. (2011). *A sceptical climate: Media coverage of climate change in Australia. Part 1—Climate change policy.* Sydney: Australian Centre for Independent Journalism.

Bailey, I., MacGill, I., Passey, R., & Compston, H. (2015). The fall (and rise) of carbon pricing in Australia: A political strategy analysis of the carbon pollution reduction scheme. *Environmental Politics, 21*(5), 691–711.

Blanco, I., Lowndes, V., & Pratchett, L. (2011). Policy networks and governance networks: Towards greater conceptual clarity. *Political Studies Review, 9*(3), 297–308.

Boykoff, M. T., & Boykoff, J. M. (2004). Balance as bias: Global warming and the US prestige press. *Global Environmental Change, 14*(2), 125–136.

Bulkeley, H. (2000). Discourse coalitions and the Australian climate change policy network. *Environment and Planning C: Government and Policy, 18*(6), 727–748.

Chilvers, J., & Evans, J. (2009). Understanding networks at the science-policy interface. *Geoforum, 40*(3), 355–362.

Chubb, P. (2012). Really, fundamentally wrong: Media coverage of the business campaign against the Australian carbon tax. In E. Eide & R. Kunelius (Eds.), *Media meets climate: The global challenge for journalism* (pp. 179–194). Gothenburg: Nordicom.

Chubb, P. (2014). *Power failure: The inside story of climate politics under Rudd and Gillard*. Melbourne: Black Inc.

Chubb, P., & Bacon, W. (2010). Australia: Fiery politics and extreme events. In E. Eide, R. Kunelius, & V. Kumpu (Eds.), *Global climate, local journalisms: A transnational study of how media make sense of climate summits* (pp. 51–65). Freiberg: Projekt Verlag.

Comby, J. (2015). *La question climatique: sociologie d'un processus de dépolitisation*. Paris: Raisons d'Agir, coll. Cours & Travaux.

Compston, H. (2009). Networks, resources, political strategy and climate policy. *Environmental Politics, 18*(5), 727–746.

Feldman, O. (2011). Reporting with wolves: Pack journalism and the dissemination of political information. In T. Inoguchi & J. Purnendra (Eds.), *Japanese politics today: From Karaoke to Kabuki democracy* (pp. 183–200). New York: Palgrave Macmillan.

Flew, T. (2013). FactCheck: Does Murdoch own 70% of newspapers in Australia? *The Conversation*, August 7. Retrieved from https://theconversation.com/factcheck-does-murdoch-own-70-of-newspapers-in-australia-16812

Hallin, D. C., & Mancini, P. (2004). *Comparing media systems: Three models of media and politics*. New York: Cambridge University Press.

Hannam, P. (2014). Australia bets on coal as climate policy crumbles. *Sydney Morning Herald*, July 26. Retrieved from http://www.smh.com.au/environment/climate-change/australia-bets-on-coal-as-climate-policy-crumbles-20140725-zufio.html

Hulme, M. (2009). *Why we disagree about climate change*. Cambridge: Cambridge University Press.

Kingston, J. (2014). Japan's nuclear village: Power and resilience. In J. Kingston (Ed.), *Critical issues in contemporary Japan* (pp. 107–119). New York: Routledge.

Kuramochi, T. (2015). Review of energy and climate policy developments in Japan before and after Fukushima. *Renewable & Sustainable Energy Reviews, 43*, 1320–1332.

Lacroix, V., & Zaccaï, E. (2010). Quarante ans de politique environnementale en France: Évolutions, avancées, constante. *Revue Française D'administration Publique, 134*(2), 205–232.

Lascoumes, P. (1994). *L'Éco-pouvoir: Environnements et politiques*. Paris: La Découverte.

Manne, R. (2012). A dark victory: How vested interests defeated climate science. *The Monthly*, August. Retrieved from https://www.themonthly.com.au/issue/2012/august/1344299325/robert-manne/dark-victory

McKnight, D. (2012). *Rupert Murdoch: An investigation of political power.* Sydney: Allen and Unwin.

Nisbet, M. C., & Fahy, D. (2015). The need for knowledge-based journalism in politicized science debates. *The Annals of the American Academy of Political and Social Science, 658*(1), 223–234.

Olausson, U., & Berglez, P. (2014). Media and climate change: Four long-standing research challenges revisited. *Environmental Communication, 8*(2), 249–265.

Ollitrault, S. (2008). *Militer pour la planète : Sociologie des écologistes.* Rennes: Presses Universitaires de Rennes.

Rayner, S. (2006). Wicked problems: Clumsy solutions—Diagnoses and prescriptions for environmental ills. Jack Beale Memorial Lecture on Global Environment, Sydney, Australia, July. Retrieved from www.insis.ox.ac.uk/fileadmin/InSIS/Publications/Rayner_-_jackbeale-lecture.pdf

Schreurs, M. (2002). *Environmental policy in Japan, Germany, and the United States.* Cambridge: Cambridge University Press.

Skea, J., Lechtenböhmer, S., & Asuka, J. (2013). Climate policies after Fukushima: Three views. *Climate Policy, 13*(Suppl. 1), 36–54.

Taylor, L. (2011). Coalition U-turn on coal power station closures. *Sydney Morning Herald*, July 21. Retrieved from http://www.smh.com.au/environment/climate-change/coalition-uturn-on-coal-power-station-closures-20110720-1hox1.html

Tiberghien, Y., & Schreurs, M. A. (2007). High noon in Japan: Embedded symbolism and post-2001 Kyoto Protocol politics. *Global Environmental Politics, 7*(4), 70–91.

Vivoda, V., & Graetz, G. (2014). Nuclear policy and regulation in Japan after Fukushima: Navigating the crisis. *Journal of Contemporary Asia, 45*(3), 1–20.

CHAPTER 9

Following the Tweets: What Happened to the IPCC AR5 Synthesis Report on Twitter?

Dmitry Yagodin, Matthew Tegelberg, Débora Medeiros,
and Adrienne Russell

In late October 2014, the IPCC gathered at the Tivoli Congress Center in Copenhagen to complete the final stages of approval and adoption of the Synthesis Report (SYR) of the Fifth Assessment Report. After the meeting, the panel issued a press release[1] announcing a public presentation of the scientific report. The press release, officially titled "IPCC Media Advisory",

---

[1] IPCC Media Advisory. (2014, October 30). Webcast and live feed of IPCC press conference. *Intergovernmental Panel on Climate Change*. Retrieved from http://www.ipcc.ch/pdf/press/141030_webcast_advisory.pdf

---

D. Yagodin (✉)
School of Communication, Media and Theatre, University of Tampere, Tampere, Finland

M. Tegelberg
Department of Social Science, York University, Toronto, Canada

D. Medeiros
Institute for Communication and Media Studies, Free University of Berlin, Berlin, Germany

A. Russell
University of Denver, Denver, USA

© The Author(s) 2017
R. Kunelius et al. (eds.), *Media and Global Climate Knowledge*,
DOI 10.1057/978-1-137-52321-1_9

is a two-page document which outlines who will present the report, when and where. The IPCC event was scheduled on Sunday, November 2 at 11:00 a.m., Copenhagen time. IPCC Chairman Rajendra Pachauri, UN Secretary-General Ban Ki-moon and incoming COP President, Peruvian Environment Minister, Manuel Pulgar-Vidal were named as presenters. The document also provides instructions for media accreditations, technical details for connecting to the broadcast signal, and offers a hyperlink to a live webcast. The press release invites journalists to interview IPCC authors after the conference. An in-text hyperlink, clickable but also spelled out at full length, links to another IPCC document that explains how to organize such an interview. The second document, in turn, links to an online form that journalists must complete to request an interview. Finally, the original press release ends with an invitation to follow the upcoming event on Facebook, LinkedIn, and Twitter by clicking on one of the three icons.

This chapter focuses specifically on the role of Twitter—currently the dominant microblogging tool—to explore what happened to SYR from the viewpoint of social media. From this perspective, the press release described above is only a small artefact of science communication within the vast scale of global media attention paid to the IPCC and climate change. Yet this artefact can be used as an empirical entry point that can lead to fuller understanding of the dynamic new media environment and actor-networks in which IPCC communication takes place. Departing from this small document the chapter probes into the network of digital communication and analyzes how the scientific knowledge of SYR translated into a complex system of transnational stakeholders and their local sub-networks. The IPCC's new communication strategy raises an important theoretical question: does the use of social media call for reconsideration of the role of professional journalism in science communication? (Brumfiel 2009; Secko et al. 2013)

SYR as a Communication Actor-Network

What are the characteristics of the social media communication networks initiated by the SYR launch? How was the report broken into pieces and distributed in an unfolding flow of digital communication? What does the SYR look like after it is *translated* by a contingent network of social actors and modern technological tools? How can we make sense of this interplay of technological and social agencies to better understand the work of new media tools? This chapter addresses these questions with the help of material semiotics inspired by actor-network theory (Law 2009; Latour 2005), contributing to a growing literature on climate change communication in social media (Segerberg and Bennett 2011; Pearce et al. 2014; Hickman 2015; O'Neill et al. 2015; Williams et al. 2015). It takes its theoretical grounding, inspiration and central premises from actor-network theory's (ANT) call to "follow the actors" in order to find out what types of constellations of relations they form.

ANT views any social phenomena as consisting of heterogeneous relations where humans and non-humans alike are treated as actors and ascribed agency (Callon 1986; Law 1992; Latour 2005). While an established method of social scientific inquiry, the concepts of ATN have only recently been taken up by media scholars to study relations between actors in communication networks (Couldry 2008; Besel 2011; Van Den Eede 2013). Yet, as John Law (1992: p. 382) pointed out long before the social media breakthrough, non-human actants are central protagonists in any communication network:

> [O]ur communication with one another is mediated by a network of objects—the computer, the paper, the printing press. And it is also mediated by a network of objects-and-people, such as the postal system. The argument is that these various networks participate in the social.

From the viewpoint of actor-network theory, the IPCC press release can be used as a starting point to analyze the scope and the format of digital SYR reception. From this perspective the press release, and the "original" tweet about it, are essential and meaningful parts of the report itself. They act by mobilizing (or inviting) an initial network of human and non-human actors, a network that further unfolds its visible and invisible attributes as something that can be called a communication actor-network.

By "communication actor-network", we refer to associations between actors, both human and non-human, that combine to form a particular interconnected media environment (Couldry 2008). This network includes the contributions of human actors, such as IPCC officials, journalists, politicians and ordinary people who produce textual representations of the SYR. Furthermore, it encompasses an array of technological mediums used to communicate the SYR, including newspapers, websites, webcasts, and Twitter accounts. By studying associations between actors within this heterogeneous network, we aim at mapping a constellation that simultaneously constitutes what the SYR is on the public agenda and also defines the different roles that "actants" play in setting this agenda (Latour 2005).

Any moment (or singular tweet) in the networks we examine can be seen as an act of *translating* a particular version of the story. Callon explains the concept of translation as a "...mechanism by which the social and natural worlds progressively take form. The result is a situation in which certain entities control others" (1986: p. 81). Each tweet, then, rewrites the story by referring to specific sources, claims, documents, or specific arguments. Just as each news story about the SYR launch can be seen as a collection of quoted actors, statements, media formats used in the story (graphs, audio, video) and newsroom divisions (political or environmental desk). These are acts of translation because each new story, or tweet, "repeats" the original construction (a media event staged and coordinated by the IPCC) while at the same time re-contextualizing it. ANT distinguishes between different types of translation strategies (Law 1992: p. 387). Translation can influence an actant's *mobility* by ordering their relations through space, as journalists do by choosing to interview some scientists while disregarding others. Another strategy influences the *durability* of actor-networks by ordering them through temporal connections, as in narratives that emphasize the continuity of IPCC work by referring to its previous reports.

When IPCC tweeted "Watch the #IPCC webcast on Sunday #Climate2014 #ClimateChange #AR5", the four hashtags defined key themes for dissemination. By retweeting this message, adding more hashtags and hyperlinks, the global collective Twitter actor becomes part of the communication network. This is a key factor in how Twitter works: it not only offers audiences a chance to filter contents from the web, but also act as part of that filtering network. Together hashtags, hyperlinks and @mentions are intrinsic parts of the functional grammar of Twitter's network logic (Johnson 2014). Simple tweets—without hashtags—only reach immediate followers. A hashtag, however, automatically ascribes a tweet to a thread of all tweets with the same hashtag. Thus, the hashtag sign as a material property of the Twitter actor elevates a simple message to a new level of networked communication and creates the potential for wider information dissemination. Hyperlinks extend the communication network of a tweet in a different way. Adding a hyperlink to a tweet helps to overcome Twitter's 140 character limit and lead the audience to more in-depth or multimodal content, such as longer texts or visuals elsewhere on the Internet. Finally, @mentions can bring actors into "open forums" where dialogue between competing perspectives is possible (Williams et al. 2015). Different combinations of hashtags, hyperlinks and @mentions thus become building blocks of different communities of reception, interpretation and further dissemination of digitally packaged media content.

THE PROMINENCE OF DOMESTIC POLITICS AND CIVIL SOCIETY ACTORS

What appeared on Twitter a few days before, during and after the SYR press conference? What kind of tagging and hyperlinking was involved and how much? Using tools for gathering Twitter data[2] and applying some basic filtering to the search results,[3] we collected a sample of close to 40,000 tweets. The tweeting peaked on Sunday, November 2, and on the following day, generating at least ten times more IPCC-related content than during the days before or after. This short-range jump in intensity confirms other studies of Twitter use in comparable settings (O'Neill et al. 2015).

Table 9.1 lists the three main categories of actants in the SYR communication actor-network: hashtags,[4] hyperlinks (see Appendix for full source details) and @mentions.

Table 9.1 Summary of top-ranking actants in IPCC SYR Twitter data

Top 10 hashtags	%	Top 10 @mentions	%	Top 10 sources	%
#cdnpoli	2.51	@climatewwf	10.3	WWF Climate Energy blog	20.7
#climateaction	1.82	@climatesavers	10.3	BBC News	4.8
#tarsands	1.8	@ipcc_ch	3.96	IPCC's SYR	4.12
#un	1.25	@greenpeace	2.56	The Guardian 1	2.76
#actonclimate	1.23	@mikehudema	1.94	Grist.org	1.95
#cop	0.83	@bbcbreaking	1.79	WWF Global 1	1.6
#globalwarming	0.77	@wwf	1.66	The Guardian 2	1.2
#savethearctic	0.67	@irinagreenvoice	1.45	WWF Global 2	1.1
#auspol	0.65	@350	1.4	The New York Times	1.08
#divest	0.62	@grist	1.3	The Washington Post	1

[2] In this study, we used NodeXL software that helps with gathering, visualizing and analysing data from different social media platforms.

[3] The search time frame was limited to tweets published between October 30 and November 5, 2014. The resulting sample consists of 39,103 tweets, involving 18,855 unique users. Retrieving all the tweets that mention IPCC, without the hashtag sign, produces a data set approximately four times larger. Collecting such a large volume of material misses the point of tracing the active tagging and sharing practices that enrich and intensify digital communication. The total number of hashtags used in the sample is 80,937 (39,103 of them are #ipcc), the total number of unique hashtags is 1232.

[4] The list naturally excludes #ipcc. It also excludes four other frequently used hashtags that were considered thematically obvious and predictable: #climate (35.89%), #climatechange (11.61%), #climate2014 (8.75%), #ar (8.65%), and #cambioclimatico (0.88%) (Spanish for climate change).

Combining the top mentioned actants—hashtags, twitter accounts, and hyperlinked stories—highlights several distinctive characteristics of the SYR communication actor-network. The most striking result is that 8080 (21 percent) hyperlinks in the network linked to a single *WWF Cool Planet Blog* post[5], and this occurred through the centralizing capacity of Twitter. The story, "How the IPCC quieted climate sceptics with its new report," is written by Stephan Singer, Director for Global Energy Policy at World Wide Fund (WWF) International. Two additional WWF stories are included in the list, demonstrating how much retweeting activity occurred around the WWF's primary Twitter accounts: @climatewwf, @wwf and @climatesavers, a WWF program that targets industry and business.

Moreover, in the data, there is an overall prominence of NGOs (@greenpeace, @350) and other individual civil society voices, which demonstrates what type of stakeholders are more eager to translate their activities through social media. For example, Canadian Greenpeace leader Mike Hudema is a crucial contributor to the high trending positions of two specifically Canadian hashtags: #cdnpoli and #tarsands. These hashtags refer to domestic politics and carbon-intensive oil industry activities in the province of Alberta. Domestic political issues also played a powerful trend-setting role in other parts of the SYR Twitter network. The high placement of #auspol reflects the heated and politicized climate change debate in Australia. While some of these findings were predictable, others were surprising including the high frequency of mentions for online environmental activist Irina Tikhomirova (@irinagreenvoice). Tikhomirova campaigns for the protection of the Arctic (#savethearctic) by publishing her visual art projects online. Her place among the central nodes of the actor-network is evidence of the important role artistic, creative and visual cyberculture can play in social media discussions about climate science and policies. This finding confirms a common assumption that to gain visibility "…all actors must deal with the media's interest in spectacle" (Gamson and Wolfsfeld 1993: p. 125), and visual tools suit this purpose perfectly.

In Table 9.1 professional media organizations are also visible, representing the second largest group of actors in the network. This can be thought of as a relatively moderate performance. The category of top @ mentions includes *BBC* and *Grist.org*, a website that features environmental journalism and commentary. However, there is more evidence of pro-

---

[5] Singer, S. (2014). How the IPCC quieted climate sceptics with its new report. *WWF Cool Planet Blog*, November 3. Retrieved from http://climate-energy.blogs.panda.org/2014/11/03/ipcc/.

fessional media presence in the form of hyperlinked texts that circulated in the network. Apart from *BBC* and *Grist*, traditional media stories by *The Guardian*, *The New York Times* and *The Washington Post* reached the top of the sources list. It is worth noting that in the #IPCC network a specialized online environmental magazine like *Grist* was able to successfully compete with its more established journalistic counterparts.

Finally, the role of the IPCC remained considerable during the studied period. The official IPCC Twitter account and the digital copy of the SYR text[6] stayed firmly on the networked agenda with more than 4 percent of tweets linking to it. This is not surprising, given the IPCC's explicit call via its invitation to follow the press conference on social media. At least quantitatively, the point of entry to the communication actor-network served this purpose and did not lose much of its representational power as a vast and diverse web of actors joined the process.

Interpretive Communities in the #IPCC SYR Communication Network

During the SYR launch, the Twitter actants described above formed a complex communicative structure, a network of relations that can be visualized for further exploration. Figure 9.1 maps a *partial* picture of that network and reveals some central communities that were formed. Just as the most frequent hashtags or hyperlinked materials are shown in Table 9.1, only the most influential actors are intentionally left visible on the map. The category of influence is a measure defined with a NodeXL ranking algorithm similar to Google Search PageRank. The algorithm takes into account the number of incoming links (mentions, replies, retweets) that each account receives and the quality of these links. This means being @mentioned by a high profile user generates more network influence than @mentions coming from a diffuse set of occasional, low-profile users. In other words, this algorithm filters out "clicktivism" and "click farming", phenomena which can create the illusion of genuine interest and participation by social media users (Morozov 2009; Karpf 2010; Chu et al. 2012). In Fig. 9.1, the resulting measure of influence for each actor is represented by the size of the individual actors' node. We can see that influence of

---

[6] Climate Change 2014: Synthesis report. (2014). *Intergovernmental Panel on Climate Change*. Retrieved from https://www.ipcc.ch/pdf/assessment-report/ar5/syr/SYR_AR5_FINAL_full.pdf

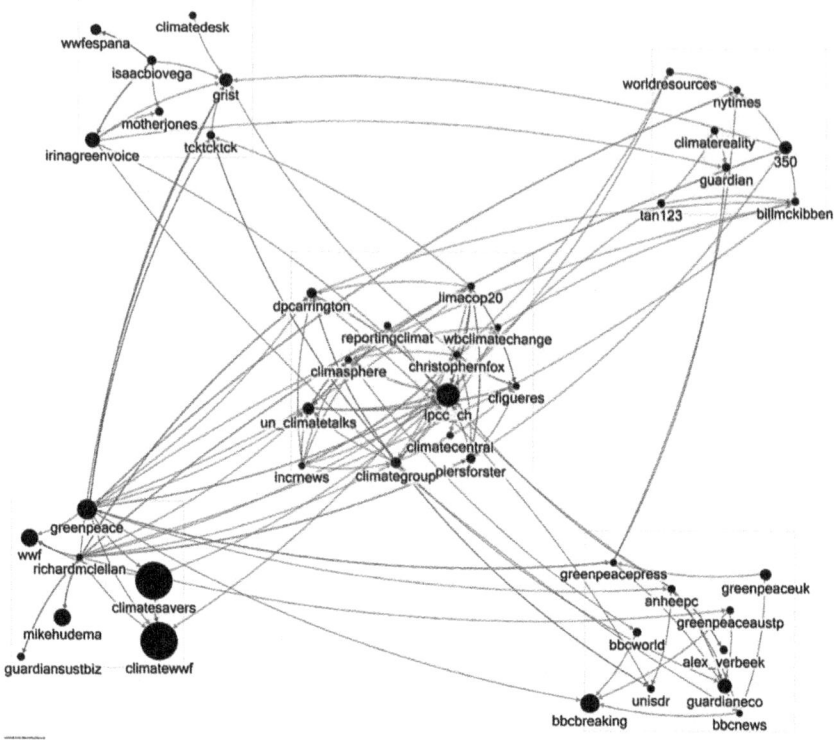

Fig. 9.1 Map of the #IPCC SYR communication actor-network

most of the actors on the map corresponds to the high frequencies of mentions and external hyperlinks we saw above (Table 9.1).

The map also represents the existing distances between individual actors and between communities. The further apart two actors are, the less likely they mention one another and the less likely their respective communities (thousands of users, invisible on the map, whose tweeting affects the overall formation) overlap. Formation of the ties occurs when tweets contain specialized @mentions that refer to other Twitter users. By applying one of NodeXL's cluster analysis algorithms (Clauset et al. 2004) we can break the SYR communication network into five distinctive communities.

There is, of course, no short and simple explanation for such divisions. However, some group dynamics are intuitively clear. This is the case for closely positioned ClimateSavers, WWF, and Richard McLellan, former director of the sustainable economy program at WWF International and current leader of the Northern Agricultural Catchments Council, an Australian environmental organization. Mike Hudema's role at Greenpeace Canada makes his placement among the same group of NGOs logical. Interestingly, a *Guardian* team that focuses on sustainability and the environmental impacts of business (@guardiansustbiz) gravitates toward the same community, reflecting the alliance within the Climate Desk, an international journalistic collaboration project. The newspaper's general environment section (@guardianeco), however, appears in the lower-right corner of the map among a cluster of UK- and Australia-based NGOs, along with several BBC accounts.

The diversifying profile of *The Guardian* reveals itself through the closeness of the newspaper's main Twitter account (@guardian) to the US-based movement *350.org* and its founder Bill McKibben. The same community, in the upper-right corner of the map, includes such actors as *The New York Times*, the World Resources Institute, a non-profit based in Washington, and well-known climate skeptic Tom Nelson (@tan123). The head of *The Guardian* environment section, Damian Carrington (@dpcarrington), is part of a network community centered on the IPCC. This community (also central on the map), despite the visually low influence of its parts, displays a very dense net of connections between the IPCC and various related actors, such as Christiana Figueres from UNFCCC, climate scientists Piers Forster and Christopher N. Fox, and a feed for what was then an upcoming COP conference in Lima, Peru. Proximity to such initiatives as the Climate Group, a non-profit alliance of corporations and authorities at different national contexts, or Climate Central, a climate science and journalism research organization, strengthens the institutional character of this interpretive community and, thus, differentiates it from the others.

The upper-left corner of the map shows a point of contact to non-English sub-networks with an example of the Spanish WWF representative Isaac Vega (@isaacBIOvega). Vega uses his personal feed to build connections between his place of work (represented by @wwfespana) and a number of established climate change news websites. Hence, he plays the role of translating (both linguistically and in actor-network terms) the IPCC SYR as a media event to the Spanish-speaking community, which represented the second largest language group in the network. For the most part, he does this by proxy, referring to climate news sites such as *Grist* and *MotherJones*, but also by commenting on and linking to environmental art projects and tweets by Irina Tikhomirova (@IrinaGreenVoice).

THE CRITICAL ADVOCACY OF NGOs

In our Twitter sample, the media resources of World Wide Fund (WWF) prevailed in disseminating information about the SYR's findings. Yet this global environmental organization does not quote from or refer to the IPCC officials who presented the report. Instead, WWF uses their own staff to comment on the findings. From the viewpoint of Twitter communication, therefore, it is Stephan Singer, director for Global Energy Policy for WWF International, and Samantha Smith, leader of WWF's Global Climate and Energy Initiative and their texts that become central actants in this communication network. In "How the IPCC quieted climate sceptics with its new report" (Singer) and "No more debates on climate science, over to leaders" (Smith), WWF picks up the SYR story and dissolves the intergovernmental diplomatic authority of IPCC language into its own environmentalist agenda wherein "...people from all parts of society marched to demand action"[7]. This translation from the intergovernmental realm into the realm of non-governmental civil society groups emphasizes politicized calls to "act swiftly and with ambition". This translation, then, crosses fluently from a descriptive scientific authority into a prescriptive one, something the IPCC carefully tries to avoid in its own messaging (Chap. 2).

The activism-driven interpretation of science comes to be molded in (and multiplied by thousands of retweets) catchphrases like "debunking global cooling" and "quieted climate sceptics", simplifying the climate science to the level of a heated discussion between opponents. However, in the WWF texts, there are only vague references to climate skeptics, with indirect quotes of their arguments or descriptions of the media attention they received before the SYR was released. No one in particular is named, and there are no pronounced ties with those excluded. This is an act of *translation* which establishes the limits of an actor-network by limiting the *mobility* of potential actants (Law 1992), or participants in Twitter communication. The nameless references in WWF texts are used to reject

[7]Singer, S. (2014). How the IPCC quieted climate sceptics with its new report. *WWF Cool Planet Blog*, November 3. Retrieved from http://climate-energy.blogs. panda.org/2014/11/03/ipcc/; WWF Global. (2014, November 2). No more debates on climate science, over to leaders. *wwf.panda.org*. Retrieved from http:// wwf.panda.org/what_we_do/footprint/climate_carbon_energy/?232090/ No-more-debates-on-climate-science-over-to-leaders

climate skeptics in general and emphasize that the IPCC findings are solid. Instead of expanding the IPCC SYR connections in space, where other entities and potential realms can be attached, WWF actors increase the network *durability* by referring to the 1996 IPCC AR2, a report that provided the first evidence of the human impact on the climate.

All three WWF texts praise the SYR, stressing its accuracy and firm standing within the scientific community. They use the findings to highlight the urgency for governments to take action to stop climate change. While the two articles on WWF's global website[8] resemble press releases, opening with the SYR launch event in Copenhagen, the *WWF Cool Planet Blog* post by Singer offers a thorough contextualization of the scientific findings of the report. It also underlines the nature of the IPCC as an intergovernmental organization permeated by diverging interests defended by each government. In particular, the process by which scientists are selected and the contrast between what the report presents and the claims made by climate skeptics (labeled as "junk science and junk economy" at the end of the blog post). WWF also points to some of the report's limitations and indicates where it, as an NGO, diverges from it:

> They do not represent all of what we as WWF want or promote. For instance, being fully "technology-neutral", the IPCC will never promote 100% renewables by 2050. Still, even with this caveat, its overall scientific rigour and vigour is leading the world community to do what it ought to do[9].

Despite the lack of attention to the SYR presenters, the WWF's attitude toward the IPCC is respectful, almost reverent. WWF makes it no secret that the report backs up its own agenda as an organization, which helps explain why they place stress on the IPCC's credibility and the robustness of its findings.

---

[8]WWF Global. (2014, October 27). IPCC report casts halo over UN climate negotiations. *wwf.panda.org*. Retrieved from http://wwf.panda.org/wwf_news/press_releases/?231734/IPCC-report-climate; WWF Global. (2014, November 2). No more debates on climate science, over to leaders. *wwf.panda.org*. Retrieved from http://wwf.panda.org/what_we_do/footprint/climate_carbon_energy/?232090/No-more-debates-on-climate-science-over-to-leaders

[9]Singer, S. (2014). How the IPCC quieted climate sceptics with its new report. *WWF Cool Planet Blog*, November 3. Retrieved from http://climate-energy.blogs.panda.org/2014/11/03/ipcc/

Mass Media Domestication

The top tweeted traditional journalistic articles in *The New York Times* and *The Washington Post* frequently quote the IPCC report and presenters Rajendra Pachauri and Ban Ki-moon. Justin Gillis' article for *The New York Times* sums up the report's central findings and puts them in national context by quoting President Barack Obama[10]. The journalist *translates* the SYR, pulling out key points and quotes from the report itself and combining them with references to how the US political leader responded, taking the report seriously, while also emphasizing the energy industry's lack of acknowledgement or action. Writing for *The Washington Post*, Joby Warrick and Chris Mooney establish distance from the IPCC report and its authors by putting quotation marks around key facts (e.g. we are locked into an "irreversible" course), leaving room for readers to question the veracity of the report[11].

Among the US-based media sources, John Light's story for *Grist* featured the most direct quotations from individual scientists[12]. In this respect, the *Grist* coverage stands out when compared to two newspaper stories that quoted the report more often than the scientists. The story used strong warning language and referred to the report, as well as past IPCC reports, as evidence rather than point of view. In the same manner as WWF, the story used stronger warning language and referred to the report, as well as past IPCC reports, as evidence rather than point of view. The story translates the IPCC report into the "what you need to know" form of a user's guide. Symptomatically, all three US media sources displayed no contesting quotes or points of view that criticized the report.

Like traditional journalistic media in the United States, British mass media paid considerable attention to the SYR presenters. *BBC* environmental correspondent Matt McGrath and *The Guardian's* Damian Carrington both start their much re-tweeted stories by quoting the central

---

[10] Gillis, J. (2014). U.N. panel issues its starkest warning yet on global warming. *The New York Times*, November 3. Retrieved from http://www.nytimes.com/2014/11/03/world/europe/global-warming-un-intergovernmental-panel-on-climate-change.html

[11] Warrick, J., & Mooney, C. (2014). Effects of climate change "irreversible," U.N. panel warns in report. *The Washington Post*, November 2. Retrieved from http://www.washingtonpost.com/national/health-science/effects-of-climate-change-irreversible-un-panel-warns-in-report/2014/11/01/2d49aeec-6142-11e4-8b9e-2ccdac31a031_story.html

[12] Light, J. (2014, November 2). The 10 things you need to know from the new IPCC climate report. *Grist.org*. Retrieved from http://grist.org/climate-energy/the-10-things-you-need-to-know-from-the-new-ipcc-climate-report

figure at the SYR press conference, UN Secretary-General Ban Ki-moon[13]. The key message that the journalists adopted from the Secretary's press conference speech was that "Science has spoken. There is no ambiguity in the message. Leaders must act. Time is not on our side". Carrington also included a quote from Rajendra Pachauri which emphasizes hope and the potential for solutions to avert climate crisis. The two stories referred to a range of other actors from politics (UK energy and climate change secretary Ed Davey, and US Secretary of State John Kerry), economics (Lord Nicholas Stern), science (citing IPCC participants and outsiders commentators), NGOs (WWF, 350.org, Greenpeace, Action Aid) and other UN agencies (World Meteorological Organization).

A second *Guardian* news item, by Adam Vaughan, drew entirely on language from a draft of the SYR. It featured six graphs that represent evidence, collected by the IPCC, of humanity's contribution to global warming and its projected impacts. The story, published on October 31, two days before the press conference, announced what to expect from the report. It contained no references to the main actors who were to present the SYR, nor to any other actors. In fact, it is the only text among the top circulated materials published prior to the SYR presentation that continued to be a key source for Twitter users afterwards.

The three UK stories place emphasis on risks associated with failure to cut carbon emissions sharply in the coming years. Discussion of risks is paired with emphasis on potential solutions described in the SYR; namely that dangerous global warming can be avoided through quick and decisive (government) action.

Texts produced for mass media quoted a more diverse range of relevant actants than texts by NGOs. This included references to the NGOs that produced texts on the SYR and thereby increased their overall publicity, reinforcing their dominant role within the Twitter-based communication network. This generates an important networking effect when, due to professional norms, parts of the network (journalists) generate ties with other parts (non-journalists writing for NGOs) without reciprocity. Public attention moves from mass media to civil society actors but does not nec-

[13] McGrath,M.(2014).Fossilfuelsshouldbephasedoutby2100saysIPCC.*BBC News*,November 2. Retrieved from http://www.bbc.com/news/science-environment-29855884; Carrington, D. (2014). IPCC: Rapid carbon emission cuts vital to stop severe impact of climate change. *The Guardian*, November 2. Retrieved from http://www.theguardian.com/environment/2014/nov/02/rapid-carbon-emission-cuts-severe-impact-climate-change-ipcc-report

essarily return as it does in networks with mutual connections (alliances of different NGOs, for instance).

The IPCC's SYR is the main basis for commentary from many of the quoted actors described above. Each of these actors refers to findings in the SYR that lend support to their own interests. For example, UK politician Ed Davey draws parallels between risk language in the IPCC report and his claim that the UK is leading the world in its aggressive efforts to reduce domestic carbon emissions. The NGO spokespersons, cited in @dpcarrington's *Guardian* report, refer to the SYR to vindicate their strong calls for climate action. This is exemplified by Bill McKibben:

> For scientists, conservative by nature, to use "serious, pervasive, and irreversible" to describe the effects of climate falls just short of announcing that climate change will produce a zombie apocalypse plus random beheadings plus Ebola. "Breaking the power of the fossil fuel industry would not be easy", McKibben said. "But, thanks to the IPCC, no one will ever be able to say they weren't warned"[14].

There are several other examples of this practice of *translation* produced by actors communicating the IPCC findings for particular contexts. Two noteworthy examples, from the same *Guardian* story, link the SYR to technological solutions (carbon capture and storage) and to global climate negotiations, specifically the divide between developed and developing nations over emissions reduction targets.

MULTIMEDIA ATTRIBUTES OF THE NETWORK

What differentiates traditional media output from the content produced by NGOs is the broader use of multimedia additions by professional journalists—signs of professionalization and dominant organizational culture in traditional media newsrooms. The top BBC story, for instance, relies on rich multimedia tools to complement the coverage. The story includes three short YouTube videos. The first features an excerpt from UN Secretary-General Ban Ki-moon's live presentation at the IPCC SYR launch. Another video features Dr. Helen Czerski, and other climate scien-

---

[14]Carrington, D. (2014). IPCC: Rapid carbon emission cuts vital to stop severe impact of climate change. *The Guardian*, November 2. Retrieved from http://www.theguardian.com/environment/2014/nov/02/rapid-carbon-emission-cuts-severe-impact-climate-change-ipcc-report

tists, explaining the impact of climate change on Antarctica. A third video provides a 60-second answer to the question: "Why have oil prices been falling recently"? These three videos interweave the SYR-driven story with topics leading to completely different terrains: the global intergovernmental dialogue focusing on the IPCC media event, the locally bound consequences of climatic change, and the political economy of energy markets. By linking to the original press conference video the *BBC* provided the IPCC with the ideal media feedback it sought, with the SYR enjoying the closest possible rendering of what happened at the press conference and of how the report should be perceived.

John Light's post for *Grist* and Adam Vaughn's story for *The Guardian* provide more examples of stories that are exceptionally visual in orientation[15]. The visualizations are once again borrowed directly from official IPCC documents, emphasizing the power of digital content mobility and the potential for *translations* which strengthen the authenticity and objectivity of reporting on the one hand, and secure, at least to some extent, proper translation of the story on the other hand. And yet the complex science data that the IPCC has distilled into drawings and graphs are translated further by journalists, into a language more accessible to general audiences. *Grist* reduces the SYR to "10 things you need to know." *The Guardian* unpacks the climate science in six key graphs that refer to various aspects of the IPCC findings. Maps and figures provide an accessible summary of the complex scientific evidence that informs the SYR's conclusions concerning humanity's impact on the climate.

Unlike many other mass mediated stories, all the WWF stories rely exclusively on written texts and hyperlinking. The stories are generally shorter than those by professional journalists. Studying the intensively networked texts it follows that mass media are naturally more likely to exploit their archival materials, including hyperlinks to relevant editions and related past publications. This introduces a sub-network of internally thematic texts that are assembled for interested readers to follow. By contrast, the linking patterns in WWF stories are outward looking, with most hyperlinks leading either to original IPCC documents, or to similar sci-

[15] Light, J. (2014, November 2). The 10 things you need to know from the new IPCC climate report. *Grist.org*. Retrieved from http://grist.org/climate-energy/the-10-things-you-need-to-know-from-the-new-ipcc-climate-report; Vaughn, A. (2014). IPCC report: Six graphs that show how we're changing the world's climate. *The Guardian*, October 31. Retrieved from http://www.theguardian.com/environment/2014/oct/31/ipcc-report-six-graphs-that-show-how-were-changing-the-worlds-climate

entific reports. This difference is a sign of two different types of communication actor-networks unfolding. Mass media mobilize their internal resources and home-grown expertise to translate the SYR story according to norms and standards that they can control; this is also commercially important for keeping audiences on the same website. The WWF stories use the SYR press conference as part of a larger and lengthier campaign, reducing it to content that encourages advocacy. However, they do so by hyperlinking to the original scientific resources unlike mainstream journalists who try to process and translate this information themselves.

CONCLUSION

By inviting audiences to follow the press conference online (through social media), the IPCC initiated network communication that transcends the exclusive institutional framework of legacy media. This move increases the transparency of climate science and allows for wider public discussion in the format of "open forums" (Williams et al. 2015). It also opens up opportunities for alternative and even marginal voices to interact with the IPCC, using its message as part of their own communication activities. This new measure by the IPCC raised the theoretical question that guided this chapter: whether the growing use of social media calls for reconsideration of the role of professional journalism in science communication?

On one hand, the findings show that traditional journalistic organizations remain key disseminators of scientific knowledge to the wider public. They do so through prevalent use of official statements and quoted voices, such as the IPCC, UN and representatives of domestic political circles. At the same time, Twitter accounts expand the areas where traditional mass media can develop their presence. Global mass media organizations, The Guardian being a notable example, can diversify their social media profiles and become part of several interpretive communities that do not otherwise overlap or have very little in common. On the other hand, analysis of the SYR communication actor-network suggests the IPCC has limited control over what happens to a media story once it starts circulating through dispersed social media networks and interpretive communities. Science communication facilitated by social media translates into non-journalistic interpretations by civil society organizations that promote their own agenda instead of following the official agenda and voices. However, the clustering of Twitter accounts (Fig. 9.1) reveals that NGO actors are confined within the limits of their own interpretive communities more frequently than their mass media counterparts.

The prominence of established NGOs, like WWF, Greenpeace and 350.org, corroborates earlier findings (Gamson and Wolfsfeld 1993; McCluskey 2008) that civil society organizations who invest more in professionalizing their communication strategies become more frequent sources for mainstream media. As the connection between *The Guardian* and 350.org shows, the formation of clusters featuring mainstream media Twitter accounts together with the accounts of NGOs may be a combined result of the resources spent on strategic communication efforts (such as the Climate Desk partnership) and the level of professionalization. This is a common practice. Although one cannot measure the exact influence that NGOs have on the public agenda, there is clearly an increasing tendency for civil society organizations to engage with the media and state officials, pursuing the pragmatic interests of gaining publicity and producing more news regarding their own practices (Waisbord 2011). The interpretive communities that we observe in the case of the SYR Twitter network are largely structured by the active role of non-governmental organizations and other civil society actors. When these actors gain prominence quantitatively and qualitatively—they enjoy a substantial social media following, becoming part of traditional mass media coverage of climate change.

The chapter brings to light another important distinction. In moments of providing context, mass media *translations* of the IPCC SYR employ a strategy that increases the *mobility* of various voices that substantiate discussions on the topic in a descriptive and often uncritical manner. Dominant civil society actors, however, are prone to increase the network mobility of organizations with similar interests through a kind of environmental solidarity, with prescriptive and critical undertones. With the help of hyperlinking tools, the IPCC holds the middle ground in this translation spectrum; it produces documents that circulate unchanged and lead back to the IPCC itself. After all, was one of the IPCC's main intentions not to extend its own publicity to social media with as few "translating" intermediaries as possible?

BIBLIOGRAPHY

Besel, R. D. (2011). Opening the 'black box' of climate change science: Actor-network theory and rhetorical practice in scientific controversies. *Southern Communication Journal, 76*(2), 120–136.

Brumfiel, G. (2009). Science journalism: Supplanting the old media? *Nature, 458*, 274–277. Retrieved from http://www.nature.com/news/2009/090318/full/458274a.html

Callon, M. (1986). Some elements of a sociology of translation: Domestication of the scallops and the fishermen of Saint Brieuc Bay. In J. Law (Ed.), *Power, action and belief: A new sociology of knowledge* (pp. 196–233). London: Routledge.

Chu, Z., Widjaja, I., & Wang, H. (2012). Detecting social spam campaigns on Twitter. *Applied Cryptography and Network Security, 7341*, 455–472.

Clauset, A., Newman, M. E. J., & Moore, C. (2004). Finding community structure in very large networks. *Physical Review E, 70*, 1–6. Retrieved from http://arxiv.org/abs/cond-mat/0408187

Couldry, N. (2008). Actor network theory and media: Do they connect and on what terms? In A. Hepp, F. Krotz, S. Moores, & C. Winter (Eds.), *Connectivity, networks and flows: Conceptualizing contemporary communications* (pp. 93–110). Cresskill, NJ: Hampton Press.

Gamson, W. A., & Wolfsfeld, G. (1993). Movements and media as interacting systems. *Annals of the American Academy of Political and Social Science, 528*, 114–125.

Hickman, L. (2015). The IPCC in an age of social media. *Nature Climate Change, 5*, 284–286.

Johnson, J. (2014). Twitter. In K. Harvey (Ed.), *Encyclopedia of social media and politics* (pp. 1282–1287). Los Angeles: Sage.

Karpf, D. (2010). Online political mobilization from the advocacy group's perspective: Looking beyond clicktivism. *Policy & Internet, 2*(4), 7–41.

Latour, B. (2005). *Reassembling the social: An introduction to actor-network-theory.* London: Oxford University Press.

Law, J. (1992). Notes on the theory of the actor-network: Ordering, strategy and heterogeneity. *Systems Practice, 5*, 379–393.

Law, J. (2009). Actor network theory and material semiotics. In B. S. Turner (Ed.), *The new blackwell companion to social theory* (3rd ed., pp. 141–158). Oxford: Blackwell.

McCluskey, M. R. (2008). Activist group attributes and their influences on news portrayal. *Journalism & Mass Communication Quarterly, 85*(4), 769–784.

Morozov, E. (2009). The brave new world of slacktivism. *Foreign Policy*, May 19. Retrieved from http://foreignpolicy.com/2009/05/19/the-brave-new-world-of-slacktivism/

O'Neill, S. J., Kurz, T., Williams, H. T., Wiersma, B., & Boykoff, M. (2015). Dominant frames in legacy and social media coverage of the IPCC Fifth Assessment Report. *Nature Climate Change, 5*, 380–385.

Pearce, W., Holmberg, K., Hellsten, I., & Nerlich, B. (2014). Climate change on Twitter: Topics, communities and conversations about the 2013 IPCC Working Group 1 report. *PLoS ONE, 9*(4), 1–11.

Secko, D., Amend, E., & Friday, T. (2013). Four models of science journalism: A synthesis and practical assessment. *Journalism Practice, 7*(1), 62–80.

Segerberg, A., & Bennett, W. L. (2011). Social media and the organization of collective action: Using Twitter to explore the ecologies of two climate change protests. *Communication Review, 14*(3), 197–215.

Van den Eede, Y. (2013). Opening the media-ecological black box of Latour. *Explorations in Media Ecology, 12*(3&4), 259–266.

Waisbord, S. (2011). Can NGOs change the news? *International Journal of Communication, 5*, 142–165.

Williams, H. T. P., McMurray, J. R., Kurtz, T., & Lambert, H. F. (2015). Network analysis reveals open forums and echo chambers in social media discussions of climate change. *Global Environmental Change, 32*, 126–138.

CHAPTER 10

Climate Change and Development Journalism in the Global South

Goretti Nassanga, Elisabeth Eide, Oliver Hahn,
Mofizur Rhaman, and Billy Sarwono

Good development journalism does not serve a government. It serves the
society as a whole by doing its best in giving information on positive as
well as negative aspects of development projects.

Prangtip Daorueng (2008: p. 248).

G. Nassanga (✉)
Department of Journalism and Communication, Makerere University Kampala,
Kampala, Uganda

E. Eide
Department of Journalism and Media Studies, Oslo and Akershus University
College of Applied Sciences, Oslo, Norway

O. Hahn
Centre for Media and Communication, University of Passau, Passau, Germany

M. Rhaman
Department of Mass Communication and Journalism, University of Dhaka,
Dhaka, Bangladesh

B. Sarwono
Department of Communication Science, Universitas Indonesia,
Jakarta, Indonesia

© The Author(s) 2017   213
R. Kunelius et al. (eds.), *Media and Global Climate Knowledge*,
DOI 10.1057/978-1-137-52321-1_10

Climate change and development are intrinsically linked to each other. On one hand, climate change is understood as the consequence of past development activities around the world, particularly in the West. On the other hand, climate change poses challenges to present and future development planning and implementation throughout the world, particularly in small island states and other developing and vulnerable countries. Starting from this increasing and complex overlap of development and climate change politics, this chapter examines IPCC AR5 coverage in six developing countries (Bangladesh, Chile, Egypt, Ethiopia, Indonesia and Uganda). These six countries were among the 22 countries where researchers monitored media coverage after the release of each of the four IPCC Reports. The chapter applies and develops the concept of "development journalism" and discusses the background for the low coverage of the reports.

The notion of "developing countries" is a complex one. There are a variety of rubrics for grouping such countries including "Third world", "underdeveloped countries", "least developed countries" (LDCs)—sometimes also extending the notion to "emerging economies" or "newly industrialized countries" (see, for instance, Aycan 2002). In climate politics, this vagueness has been particularly problematic, as the category "developing countries" has been extended to cover small island states as well as vast continent-like emerging economies such as China or India. This is amply illustrated during COP negotiations by groupings such as the G77, BRICS (Chap. 7), China plus G77, and so on.

Table 10.1 shows a comparison of background data on the six countries whose AR5 coverage we focus on in this chapter. Since "development" is most often measured by GDP, or by more nuanced indexes such as the Human Development Index (HDI), it is clear that the chosen countries mostly belong to the least advantaged group.[1] The table also shows how carbon emissions and other development indicators overlap. This illustrates not only the way climate politics and adaptation issues are linked to broader development debates but also how a lower position in the development scales is linked to less direct historical responsibility for emissions and higher levels of vulnerability and risk. From the point of view of developing countries, climate change has become kind of a double trap.

---

[1] In this group of countries, Chile is a borderline case due to its considerably high per capita GDP and HDI. However, it was included to represent Latin American countries where many livelihoods depend on an agro-based economy.

Table 10.1   Socio-economic background data for studied countries

	GDP[a]	CO2[b]	MFI[c]	Web[d]	HDI[e]	CRI[f]	VAI[g]	PA[h]	PRP[i]
Bangladesh	1.03	0.34	43	6.5	0.56	20.8	40.4	73.2	98.8
Chile	15.69	5.47	23	66.5	0.82	101.3	66.3	78.1	96.6
Egypt	3.2	2.6	50.2	49.6	0.68	116.3	49.9	74.5	87.1
Ethiopia	0.52	0.08	41.8	1.9	0.44	84.2	38.3	n/a	n/a
Indonesia	3.68	1.95	40.8	15.8	0.68	72	48.4	68.9	87.7
Uganda	0.69	0.05	31.7	16.2	0.48	84.8	38.3	68.3	86.1

[a]GDP per capita in thousands of USD (International Monetary Fund 2013; Statistics Norway 2013)

[b]Carbon emissions per capita in tonnes (EDGAR 2013)

[c]Media freedom index where the lowest score indicates more freedom (Reporters Without Borders 2015)

[d]Percentage of individuals using the Internet (ITU 2013)

[e]Human development index estimates life expectancy and living standards, where the highest score represents better results (UN Development Programme 2013)

[f]Climate risk index for 1994–2013 where the lowest score means the highest risk (Germanwatch 2015)

[g]Vulnerability and adaptation index (ND-GAIN 2013), higher score indicates better preparedness to resist threats of climate change

[h]Public climate change awareness by percentage (Lee et al. 2015)

[i]Public climate change risk perception (Lee et al. 2015). Percentage of those indicating "serious" personal risk as a share of those deemed "aware"

On one hand, the need for global mitigation threatens to close the door to the "right to development", a central justice claim made by developing nations since the Bandung conference in 1955 and acknowledged in the Brundtland Report (1987). On the other hand, existing development gaps are strongly reflected in the ability of the least developed countries to address adaptation problems caused by consequences of climate change that are already a by-product of (other nations') past emissions. Facing these realities, leaders of non-industrialized or less-industrialized nations have challenged the unequal development of rich and poor nations, and during climate negotiations focused on the historical responsibility of developed nations to carry the burden of future emission mitigation. This is followed by claims that these nations need to accept simultaneously the developing nations' need for modernization, development of adaptation capacity and demands for green technology (Page 1999; Bond 2012).

Deep underlying inequities and climate injustice are well illustrated by the inner contradictions in suggesting global remedies to climate change. When richer nations provide "climate finance" in the form of loans (not grants), they are in effect asking poorer nations to pay to fix a problem richer nations themselves have created, yet the communities that are most vulnerable to climate change often have done the least to contribute to the problem (Banda 2013: p. 10).

Development Journalism

The idea of development journalism has been interpreted and practiced differently across the world. It can be traced to the early post-colonial era, linked to the movement for a New World Information and Communication Order (NWICO), as a demand made by a group of non-aligned develop-ing countries. Following concerns related to the hegemony of Western perspectives in international news flows, some Third World journalists felt the need to have more government input in the news reporting process as a means of realizing national development goals. This came to be termed development journalism (Ogan 1980) and became part of a "modern-ization paradigm" (Banda 2008: p. 155). Thus, development journalism started as a distinct journalistic practice in Asia, Africa and Latin America, with media believed to play a significant role particularly for the devel-opment of newly independent states (Moemeka 1989; Schramm 1964; Skjerdal 2011; Xu 2009). In Asia, its birth was associated with the his-tory of the Press Foundation Asia (PFA) as part of the International Press Institute (IPI).

The practice of development journalism went through a period of dependency-dissociation, which entailed opposition to the first phase's dependency on North–South technology transfer, while maintaining strong state–journalism relationships. A Freire-inspired critique of this relationship opted for new more citizen-oriented practices (Banda 2008). Sharma (2007) considers development journalism as a "vast arena cover-ing practically every subject about the human condition, from poverty to global warming". She argues that when "journalists view a story from the development angle and connect it to people's lives, they are engaged in development journalism" (Sharma 2007: p. 113). Thus, the emergence of this brand of journalism saw a contestation developing between state pro-motion and views from below. Authoritarian regimes promoted develop-ment journalism based on the belief that journalism should cooperate with governments in nation building and overall social, economic and political development (Xu 2009: p. 362).

This perspective on development journalism is a premise of McQuail's (2003) development media theory, where media was expected to focus on the goal of national development and thus "be careful not to undermine the authority of government and its agents through 'irresponsible' criticism" (p. 56). McQuail acknowledges that while these ideas expressed a version of social responsibility, in practice they often serve to legitimize the surrender of independence and subordination to whoever has power. This view of complacent journalism has sometimes been misused to suppress critical voices against government and limited journalists in their ability to perform a watchdog role. As Skjerdal (2011) finds in his study of Ethiopia's official policy on "development journalism", this framework has indeed prompted journalists to respect the cause of nation building, while side lining professional values of media independence. Consequently, reporters have stayed on the "safe side" by focusing on the positive aspects of policy rather than those that may attract negative reactions from their bosses (Skjerdal 2011: pp. 67–68).

Critics of this perspective on development journalism argue that "the role of journalists was not to support the state development goals without question [but instead] to make sure that the development process was accountable and transparent" (Daorueng 2008: p. 243). In other instances, development journalism has been used for ideological purposes to counter "Western" or "Northern" journalism models and their ideals of "objectivity". Thus, development journalism continues to be linked to dependency theory and the battle against Western cultural invasion by promoting national cultural values and identities (Xu 2009: p. 360). Against this complex history, it is not surprising that development journalism has drawn mixed reactions from both developing and developed countries. Within developing nations, Western journalists are criticized for protecting their own capitalist interests in the world when they attack development journalism by labeling it as "government say-so journalism". Western journalists, in turn, expressed fear that the establishment of development journalism would put an end to the freedom to collect information first-hand in foreign countries (Ogan 1980).

Summing up this history, Wimmer and Wolf (2005) classify development journalism in two broad camps. One strand is comparable to a Western style investigative journalism that demands reporting that critically examines development projects and where freedom of the press is a basic requirement. The second strand of development journalism is a more benevolent–authoritarian approach that allows for systematic manipulation of information in favor of a subtle development serving common welfare. In actual practice these two categories are not mutually exclusive as journalists in most countries select elements from both. The normative power associated with the concept of "development" is a common denominator in both versions.

George (2009) draws on *The Guardian's* Katine development project in Eastern Uganda to define development journalism. She contends that "… we are looking at big policies affecting developing countries and looking at how this relates on the ground to those who expect to be benefiting". She explains that while the Ugandan government may have a poverty-reduction strategy in place, the interest is to show what this actually means for people living in Katine village.[2] In practical terms, one of the journalists reporting on climate change for the *Daily Monitor* newspaper disclosed that he considers himself a development journalist. What motivates his reporting is the desire to contribute to improving people's welfare and have a positive impact on their lives. In his words, development journalism aims at "improving people's life through reporting about issues on health, environment, agriculture, business and other areas of social development" (Emmanuel Ainebyona, personal communication, 24 April 2015).

Fackson Banda argues that development journalism must also address audiences not as consumers, but as citizens. This approach demands that journalists engage in listening to citizens' concerns and facilitate deliberation among people, as well as between people and their leaders. The goal of this, he says, is to allow citizens to frame their own development concerns. This also implies that journalism is not neutral, but develops solidarity with people; thus being both engaged and engaging (Banda 2008). As digitalization and globalization rapidly change media environments, these targets become more complex but retain their importance. As Daorueng (2008) argues:

> […] because poverty continues to pose a threat to millions of people, journalists have no choice but to continue reporting development issues and its impact on people. Development journalism of today must be able to examine and explain […] the complexities of the development process in order to help societies react sensibly to challenging global situations (p. 246).

---

[2] Katine is a sub-county in the Soroti District of Uganda. It consists of 66 villages, with a development project funded by African Medical and Research Foundation (AMREF) and Farm-Africa.

What does AR5 coverage in developing countries look like if we analyze it against this rich and sometimes contradictory legacy of development journalism? How do journalists in developing countries frame the issue of climate change as a broad development question and at the same time keep their journalism people-oriented and thus relevant to local audiences? We now turn to a more detailed analysis in six countries by tackling three themes: the amount of coverage, the particular characteristics of the coverage and the articulation of justice themes in the coverage.

WHY LOW COVERAGE IN THE "GLOBAL SOUTH"?

The amount of attention that news outlets in developing countries paid to the AR5 was, globally speaking, very low (Chap. 4). Accordingly, the first thing to note about coverage in developing countries is this general incapacity of newspapers to bring the IPCC climate science under broad local public scrutiny. This observation is consistent with findings from a previous study (Zamith et al. 2012: p. 334) which noted a lack of focus on scientific discourse from non-US media. We will discuss the reasons for this inability more extensively below. To briefly put this into perspective, the combined number of stories produced by the six developing countries in this sample was 113, which is fewer than the total number of stories in Finland alone (147). Table 10.2 shows how this (limited) volume of coverage corresponded with the launch dates for contributions from each of the IPCC Working Groups (WGs).

Bangladeshi newspapers provided the most coverage with 33 stories, while coverage in Egypt was the lowest with only 5 stories. Over the four launch periods, the Synthesis Report (SYR) generated the most atten-

Table 10.2 Number of stories by Working Group (WG) from 6 countries in the Global South

	WGI	*WGII*	*WGIII*	*SYR*	*Total*
Bangladesh	4	2	3	24	33
Chile	10	6	2	8	26
Egypt	2	3	0	0	5
Ethiopia	1	2	10	0	13
Indonesia	3	3	1	7	14
Uganda	4	1	5	12	22
Total	24	17	21	51	113

tion accounting for 45 percent (51) of the articles. The same pattern was reflected at country level with the SYR representing 73 percent of the total coverage in Bangladesh, 55 percent in Uganda and 50 percent in Indonesia. Egypt and Ethiopia, however, had no coverage of the SYR, while in Chile WGI had the most coverage with 10 stories (39 percent). Acknowledging these variances, the analysis that follows considers some of the factors that help account for the low volume of AR5 media coverage in developing countries.

In many developing countries, journalistic resources are scant and thus coverage of the background on climate science and the perils of climate change suffers. An analysis of climate coverage in the Solomon Islands (Isbrekken 2013) demonstrates that limited journalistic resources pose a problem for this vulnerable Pacific state. Among the broader sample of developing countries examined for this research (Chap. 4), Brazil was the only nation with a significant number of stories on the AR5 (168, including online stories). Across our sample for this chapter, we noted a large dependency on agency stories in the coverage of climate change in general and the IPCC reports in particular. In the Indonesian coverage, for example, 24 stories (86 percent) are sourced from international news agencies, which generally reflect global or Northern perspectives. In addition, most of the voices quoted are IPCC authors or other international academics. Very few are Indonesians, save a group of government officials who are mandated to handle climate change programs, such as the Ministry of Environment's National Council on Climate Change. This means government concerns about climate change and its impacts, such as harvest failure, deforestation, the destruction of coral reefs, etc., have not been a media priority. This dependency on foreign agencies can be explained by "media poverty", with developing countries lacking the resources necessary to facilitate journalists getting first-hand information at the IPCC launches.

Another factor contributing to low coverage is the level of climate literacy among journalists in developing countries, where climate issues are not given due priority in spite of their influence on a country's sustainable development. Uganda offers a typical example. Of the 22 stories on the AR5, only 8 (36 percent) are by local journalists. The rest originate from foreign news sources and columnists. A similar experience pertains to Bangladesh where media institutions do not seem aware of the AR5 as a resource for reporting on climate change analytically, nor do journalists have proper training on climate issues. Instead, reporting on climate change in Bangladesh is triggered by political events such as the COP summits, while other events that lay the foundation for these summits, such as the launch of the IPCC reports, are left aside.

The lack of motivation among local journalists to report on climate issues such as the IPCC reports reflects the low rank of climate stories on the media agenda in developing countries. In their study, Boykoff and Roberts, (Zamith et al. 2012: p. 339) found that coverage of climate change remained notably diminished outside of Europe and North America, while earlier studies of coverage of environmental issues in South America found that climate change coverage had not kept pace with other areas, such as crime or the economy. This view is further confirmed by several recent studies that have analyzed African media coverage of environment and climate issues. In interviews conducted with environmental journalists and editors in Uganda, it was disclosed that climate change stories "do not sell", thus they have low priority in newsrooms (Nassanga 2013). These remarks correlate with two studies on African media coverage of environmental issues which found that, in spite of an increase in quantity, there remains a lack of knowledge about climate change among journalists (Corner 2011; Shanahan 2009). Moreover, a BBC study (Deane 2009) on the effects of climate change in ten African countries[3] found that the measures proposed to address climate change and debated at international fora are barely reported in Africa. In the same vein, other researchers have added that African voices are often absent from international climate debate and decision-making (Eide and Ytterstad 2011; Toulmin 2009). UNESCO's *Climate change in Africa: A guidebook for journalists* (Banda 2013) addresses these gaps and recommends that journalists not only report on what scientists have found, but also understand the real-life implications of this research and explain its relevance to their audiences (Banda 2013: p. 70). This echoes the development journalism plea for people-centered stories with a bottom-up approach. However, the recommendation presupposes easy access to new research findings and journalists with adequate interpretational capacity.

Two added factors that explain the low coverage are proximity and the personalities involved. The IPCC launches are mostly set in far off countries (in the North) with hardly any active involvement by local/national leaders or scientists from the developing countries. Moreover, most cov-

---

[3] Democratic Republic of Congo, Ethiopia, Ghana, Kenya, Nigeria, Senegal, South Africa, Sudan, Tanzania and Uganda.

erage in the developing countries tends to be personality-oriented; if, for example, the President of a country were present at an IPCC launch event and participated in the deliberations, this would attract coverage nationally. By contrast, the low representation of scientists from developing countries is an added factor in the low volume of coverage. On the team of 71 contributors to WGI Summary for Policymakers,[4] only four were from Asia, two from South America and none from Africa. Of the 72 authors of WGII Summary for Policymakers,[5] only ten were from Asia, five from Africa and two from South America. In addition to the limited presence of national IPCC scientists from the developing countries, national governments do not pay adequate attention by playing an active role in popularizing the IPCC reports to attract local media coverage.

Low coverage can also be explained by minimal institutional investment in science reporting. In Bangladesh, when the IPCC coverage is compared with earlier COP coverage (Rhaman 2010), it becomes evident that Bangladeshi climate journalism is driven by politics. There has been considerably more coverage for COPs, since top political leaders participated at the summits. On the other hand, local politicians commenting on the IPCC reports have not attracted the same kind of media exposure. Climate change stories often compete with events high on the respective national news agendas. In Indonesia, legislative elections on 9 April 2014 collided with parts of the monitoring periods for IPCC coverage. Moreover, throughout the year considerable coverage was dedicated to presidential elections that took place on 9 July 2014.

Climate change can also be pushed into the background by national political turmoil (Egypt, Bangladesh). In Bangladesh, several events were unfolding during the study period, including parliamentary elections that generated considerable tension. Most of the coverage before and after the release of the WGI launch focused on events surrounding these elections. Another concurrent event took place during the release of the SYR on 2 November 2014, when Bangladesh's Prime Minister attended the annual UN General Assembly in the United States, which was very important for Bangladesh. There was intense interest among citizens and media in how

[4] IPCC. (2013). Summary for policymakers. *Climate Change 2013: The physical science basis*. Geneva, Switzerland: IPCC. Retrieved from http://www.climatechange2013.org/images/report/WG1AR5_SPM_FINAL.pdf

[5] IPCC. (2014). Summary for policymakers. *Climate Change 2014: Impacts, adaptation, and vulnerability*. Geneva, Switzerland: IPCC. Retrieved from https://ipcc-wg2.gov/AR5/images/uploads/WG2AR5_SPM_FINAL.pdf

Table 10.3 Story relevance to the IPCC launch and focus

	IPCC AR5 report is the main story	General climate story with reference to IPCC AR5	General climate story without reference to IPCC AR5	Other topic with reference to either IPCC or climate	Total
Bangladesh	9	10	11	3	33
Chile	10	10	5	1	26
Egypt	3	1	1	0	5
Ethiopia	1	4	8	0	13
Indonesia	12	2	0	0	14
Uganda	1	0	4	17	22
Total	35	27	30	21	113

Prime Minister Hasina would face the UN General Assembly, given her reluctance to hold an inclusive parliamentary election. Finally, in some countries, severe poverty overshadows concerns over climate change (Egypt, Ethiopia) since citizens and media are more preoccupied with issues related to meeting their basic needs. The next section analyzes how the launch of the AR5 reports was domesticated in each of the six sampled countries.

DOMESTICATING THE IPCC AR5

Table 10.3 shows the degree of relevance of the AR5 reports in the developing countries we examined. The section places emphasis on how the launch of these reports was domesticated by the respective countries. In Uganda, the newspapers studied (*New Vision,* a semi-government paper and *Daily Monitor*, a private paper), had almost the same amount of coverage with 10 and 12 stories respectively. Of the stories published during the launches only *one* concentrated on presenting the AR5 report, "Global body calls for urgency in addressing climate change", by highlighting relevant findings.[6] Addressing the African continent specifically, the story points out that "there is strong consensus that climate change will nega-

---

[6] Serunjogi, M. E. (2014). Global body calls for urgency in addressing climate change. *Daily Monitor*, November 16. Retrieved from http://www.monitor.co.ug/News/National/ Global-body-calls-for-urgency-in-addressing-climate-change/-/688334/2523364/-/ d64641z/-/index.html

tively impact food security in Africa", and thus this story addresses a major development issue on which peoples' livelihoods depend.[7]

The Indonesian papers studied are *Kompas* (national paper) and *Koran Tempo* (local paper), both privately owned. The amount of coverage in *Kompas* was three times greater than the coverage in *Koran Tempo*. The 14 articles were mostly related to the global context, with only 25 percent addressing local issues. This was evidenced by story headlines such as "The United Nations: Climate change is caused by human activities"[8], "Global warming already destroyed Australian coral reefs"[9] , and "Global warming, the peak of earth damage 2100"[10] , which all emphasize the global angle. Most of these news items were short (less than 300 words), with less than 25 percent containing more than 600 words. The length of the articles shows that climate change, and the IPCC report, were considered low-interest issues for the media.

In Bangladesh, *Daily Star* (an elite English daily) and *Prothom Alo* (a popular Bangla newspaper) published 33 articles. The bulk of them were published during the release of the SYR on 2 November 2014. *Daily Star* published almost twice as many stories as *Prothom Alo*. As a severely affected and vulnerable nation, one might have expected that Bangladeshi newspapers would have published more items, particularly during the release of WGII, which dealt with "Impact, Adaptation and Vulnerability".[11] The results, however, were the reverse with fewer stories appearing on the WGII launch than during any other launch period (Table 10.2). Nine stories appeared with by-lines but, surprisingly, they

[7]The time-lag is worth noting with this story appearing two weeks after the SYR launch on November 2, 2014.

[8]Hardoko, E. (2013). The United Nations: Climate change is caused by human activities. *Kompas*, September 27. Retrieved from http://internasional.kompas.com/read/2013/09/27/2201143/PBB.Perubahan.Iklim.Disebabkan.Ulah.Manusia.

[9]Aditya, R. (2014). Global warming already destroyed Australian Coral Reefs. *Kompas*, April 2. Retrieved from http://techno.okezone.com/read/2014/04/02/56/964135/pemanasan-global-sudah-hancurkan-terumbu-karang-australia

[10]Pertiwi, A. (2014). Global warming, the peak of earth damage 2100. *Koran Tempo*, November 3. Retrieved from http://dunia.tempo.co/read/news/2014/11/03/117619170/pemanasan-global-puncak-kerusakan-bumi-2100.

[11]As a low-lying riparian country, Bangladesh is prone to natural disasters and floods. The 2015 Climate Change and Environmental Risk Atlas, placed Bangladesh at the top of a list of ten countries "at extreme risk", cautioning that the impacts of climate change and food insecurity could lead to increased civil unrest and violence in 32 of the 198 countries assessed (Pantsios 2014).

were all penned by foreign authors. Most of the other 24 articles were produced by international agencies. Hence, in Bangladesh the journalistic contribution to IPCC reporting was limited to translating or putting an editorial touch on foreign articles. This may reflect the deep connection between climate change and development policies in Bangladesh. Climate change has been prioritized in all policy instruments in the country.

The voices quoted in the stories do not match the role Bangladesh is playing in the climate field. Among 63 quoted voices, there were no domestic national political actors and only 4 voices quoted from domestic NGOs and domestic natural science experts. All the other voices were either from foreign national political actors (mostly UN voices with 22), UN sub-organizations (8) and media (7). While the issue of climate change was more domesticated in earlier COP coverage (Rhaman 2010), voice analysis suggests the IPCC coverage has been "de-domesticated" in Bangladesh, and that journalists did not see the IPCC reports as an opportunity for development journalism.

The Chilean coverage mainly concentrated on science (WGI) and less on impacts. The two newspapers (*La Tercera* and *El Mercurio*) addressed climate change from a global perspective and concentrated more directly on the IPCC reports than newspapers in the other countries analyzed. Some articles did bring up local development issues. For instance, an article in *La Tercera* with the headline: "Less water availability will be the main climate change impact in Chile"[12] , addresses, among others, the issue of the Chilean agricultural sector, and wine production especially. The article identifies agricultural regions that will be badly affected by water shortages resulting from climate change, and highlights how this has made local economies vulnerable to climate change. An article in *El Mercurio* entitled "Western food is not healthy for the planet" partly addresses the justice theme.[13] The article criticizes rich countries and their diet of heavy meat consumption, while also pointing out that developing countries are adapting the Western consumption habits as their incomes grow. The story cites global statistics, which show that "The 15 richest countries eat 750% more meat than the 24 poorest countries". These Chilean articles

[12] Espinoza, C. (2014). Less water availability will be the main climate change impact in Chile. *La Tercera*, April 1. Retrieved from http://papeldigital.info/lt/index.html?2014040101

[13] Guzmán, L. (2014). Western food is not healthy for the planet. *El Mercurio*, November 13. Retrieved from: http://impresa.elmercurio.com/pages/LUNHomepage.aspx?BodyID=1&dt=2014-11-13&dtB=2014-11-13&dtB=13-11-2014

address production, linking climate change to people's livelihoods, and the impacts of (over)consumption, both in Chile and globally.

While coverage of the AR5 launches as a main story was minimal in Uganda, almost all the related stories were concerned with the impact of climate change on the country. Given that Uganda is an agro-based economy dependent on rain-fed agriculture with limited resources for hi-tech farming, agriculture and rainfall patterns are a central factor in the lives of Ugandans. Thus, during the four launch periods 17 out of the 22 stories could be grouped under the risk and disaster themes, with most specifically addressing rainfall variability and the reduction in agricultural production. This framing made the information relevant and helped local audiences to identify with the reports. The risk resulting from irregular rainfall summed up in an article headlined: "Global body calls for urgency in addressing climate change"[14] . The story confirms that this applies particularly to Uganda: "In Uganda, according to experts, the effects of climate change are most manifest in rainfall variability and season changes which have affected the nation seriously". Another story focuses on the threat climate change poses for coffee production (Chap. 11).[15] Studies on the impact of climate change in Uganda have shown that it has resulted in increased food insecurity; increased conflicts over land and water resources, the spread of diseases such as malaria; soil erosion; land degradation; flood damage to infrastructure, settlements and the displacement of people by landslides (Hepworth and Goulden 2008; Okello 2011; Synovate and Deutsche Welle 2007; Wiggins and Liu 2010).

Looking at coverage of climate change from a historical perspective, James Painter (2013) observes that the media has tended to favor stories emphasizing disaster over those that concentrate on the opportunities to be gained from moving to less carbon-intensive forms of industrial activity (see also Shaw 2014). While in the Global North the opportunities focus is on actions that reduce carbon emissions, in the South the focus is on actions that can be taken to mitigate and adapt to negative climate change impacts on agriculture. In the Ugandan coverage, some stories provided

[14] Serunjogi, M. E. (2014). Global body calls for urgency in addressing climate change. *Daily Monitor*, November 16. Retrieved from http://www.monitor.co.ug/News/National/Global-body-calls-for-urgency-in-addressing-climate-change/-/688334/2523364/-/d64641z/-/index.html

[15] Tajuba, P. (2014). Temperature rise threatens coffee production. *Daily Monitor*, November 11. Retrieved from http://www.monitor.co.ug/Business/Prosper/Tempreture-rise-threatens-coffee-production/-/688616/2517510/-/3p8a8q/-/index.html

opportunity frames to guide farmers on how to sustain production in spite of prolonged droughts. For example, two stories headlined "Facts about Mubuku Irrigation scheme" and "Irrigation can beat climate change"[16] focused on irrigation as a means of overcoming droughts caused by climate change. Elsewhere, opportunity stories urged people to preserve forests and plant trees as a way of minimizing climate change impacts that have already resulted in irregular rain patterns and longer droughts. Four headlines summarizing the content of these stories read: "Kadaga sounds drum for tree planting"[17]; "We must get people out of wetlands and forests"[18] ; "After losing Uganda's forest cover, we will face a water crisis"[19] ; and "Tree planting is a good investment"[20].

A group of stories on the IPCC reports in Bangladesh highlighted alarm and disaster themes as well, elaborating on the negative consequences of climate change and advocating for what should be done to mitigate the impacts. This is evident in stories titled "Alarm rings for climate system"[21] and "Use of fossil fuel must be reduced".[22] Development issues were identified in non-IPCC-related coverage around the launches including: "Micro loans can offset effects of climate change"[23] ; "Propose monitoring wing in South Asia"[24] ; and "Ensure free movement of climate migrants"[25].

[16] African Media Initiative (2013). Facts about the Mubuku irrigation scheme. *New Vision*, October 8, p. 27.

[17] Mulondo, M. (2014). Kadaga sounds drum for tree planting. *New Vision*, April 17, p. 7.

[18] Museveni, Y. K. (2014). We must get people out of wetlands and forests. *Daily Monitor*, April 1, p. 10.

[19] Rwakakumba, M. (2014). After losing Uganda's forest cover, we will face water crisis. *Daily Monitor*, April 15, p. 10.

[20] Niyonzima, G. (2014). Tree planting is good investment. *Daily Monitor*, November 10. Retrieved from http://mobile.monitor.co.ug/Tree-planting-is-good-investment/-/691260/2516528/-/format/xhtml/-/wldcdjz/-/index.html

[21] Alarm rings for climate system: UN blames mankind for global warming, calls for curbing fossil-fuel burning. *The Daily Star*, September 28, 2013, p. 16.

[22] Use of fossil fuel must be reduced. *The Prothom Alo*, November 3, 2014, p. 7.

[23] Micro loans can offset effects of climate change: Analysts. *The Daily Star*, November 16, 2014, p. B4.

[24] Propose monitoring wing in South Asia: Greens urge government. *The Daily Star*, November 13, 2014, p. 5.

[25] Ensure free movement of climate migrants: Rights bodies urge leaders. *The Daily Star*, November 16, 2014, p. 5.

CLIMATE JUSTICE AND DEVELOPMENT

While developing countries acknowledge the imperative of taking deliberate efforts for their sustainable development, by forming regional groupings, for example, the developed countries have an obligation or responsibility to assist developing countries in their mitigation and adaptation to the effects of climate change. This has to do with historical responsibilities, since developed countries emit most of the greenhouse gases and indeed have done so in the past.

Given the context of a shared environment globally and the need for "sustainable equitable development", at both the inter-generational and intra-generational levels (Brundtland Report 1987), climate justice may be seen as an integral part of development journalism, since it is closely related to North–South relations and the right to development. The issue of climate justice and the corresponding responsibility for action is often picked up at the highest political level. For instance, in a speech addressing the Second Intergovernmental Authority on Development (IGAD) Drought Disaster Resilience and Sustainability Initiative, a regional body that brings together eight member countries (Djibouti, Ethiopia, Eritrea, Kenya, Somalia, South Sudan, Sudan and Uganda), Uganda's President Yoweri Museveni pointed out that: "We are to blame for what is happening due to uncontrolled industrial policies. However, by and large, developed nations are the major polluters and should take responsibility and cut their emissions. We should collectively demand justice". However, he admits that individual countries are also to blame for what he calls "strategic mistakes", including the delay in industrialization. He explains that the implication is that: "too many people remain in agriculture, which puts a lot of pressure on the land, forcing the peasants to encroach on the forests and wetlands. [...] the low level of electrification means the peasants continue to destroy the biomass for firewood".[26] In Indonesia, several articles addressed climate justice. For example, "The Earth is still in political struggle" reflects on the negligence of developed countries, particularly when it comes to implementing agreements.[27] The story illustrates injustice committed by UNFCCC member states, citing an example

[26] Museveni, Y. K. (2014). We must get people out of wetlands and forests. *Daily Monitor*, April 1, p. 10.
[27] Laksmi, B. L. (2014). The earth is still in political struggle. *Kompas*, November 10. Retrieved from http://www.bmkg.go.id/bmkg_Pusat/Links/Ekliping_Detail.bmkg?id=b8xb3891206gp5z23814

where two days after an AR5 report launch, Australian Prime Minister Tony Abbott announced the country would continue using coal.

Given the small amount of coverage, in general, it is not surprising that instances of actualizing justice perspectives in the news coverage were not very common (see also Chap. 6). However, the examples show the obvious presence of the justice discourse in developing countries at the high political and elite levels. This at least brings the theme into the news pages, albeit to a limited extent. Consequently, due to reasons discussed above, a more bottom-up climate justice frame (and thus a more dialogic version of development journalism) was not dominant in the coverage.

Looking at the global sample of coverage (Chap. 4), it is worth noting that issues of justice and responsibility are present in newspapers in developed countries. This is well brought out by a reportage in *The Guardian*, where a strong link between extreme weather (particularly the "super-typhoon" Haiyan in 2013) and climate change is highlighted. A story headlined, "'We expect catastrophe'—Manila, the megacity on the climate frontline" starts by stating that the "IPCC report warns life in the Filipino capital and other coastal cities will get much worse in many ways as temperatures rise"[28]. The story emphasizes the need to adapt and further develop green energy sources, as the subtitle reads: "The argument for divesting from fossil fuels is becoming overwhelming". The reporter, John Vidal, cites the Head of Climate Change for the Department of Agriculture in the Philippines, who challenges the rich countries: "Rich countries tell us to become climate resilient. We agree, but who pays? We have received possibly $5m, from the World Bank for studies to adapt to climate change, and that is all". This may be read as a statement urging for climate justice; especially since the context is that the Philippines may need 20 years to recover from the immense damages caused by the typhoon. The example demonstrates that conscious efforts are also made in "Northern" newspapers to address issues of development and justice.

[28] Vidal, J. (2014). "We expect catastrophe": Manila, the megacity on the climate frontline. *The Guardian*, March 31. Retrieved from http://www.theguardian.com/environment/2014/mar/31/ipcc-climate-change-cities-manila

Conclusions: Faces and Voices

To conclude, we concede that developing countries have their own respective development journalism orientations which correspond to their particular problems. This affects coverage of the climate discourse, including the AR5 releases. Low representation and participation from developing country scientists in the process of drafting the reports may indicate that the IPCC reports often are perceived as representing the perspectives and solutions of countries in the Global North. In agro-based developing countries with few industrial activities, the major concern is the impact of climate change on agriculture and survival. The disparities in the coverage seem to stem from the climate issue tending to be viewed through different lenses in the global North and South. In developing countries, WGII, with its focus on vulnerability, had roughly the same low level of coverage as the other WGs. While high news agency dependency may serve as one explanation for the low coverage, WGII harvested ample attention in the Global North, so one would have expected equivalent coverage in the Global South.

In order to improve coverage of climate issues, it is important to have an appreciation of the potential for a transformed development journalism. This transformation can happen at three main levels: First, development journalism must become socially responsible and politically committed to climate journalism by relating messages from climate scientists to local livelihood issues, such as poverty, illiteracy, lack of food, medicine, education, and so on. In this way, development journalism can play a role in actively addressing the environmental, economic and human development concerns in urban, rural and remote areas. Second, while the overlap of development and climate agendas may help to keep climate issues on the media agenda, journalists should be aware that governments may abuse development journalism ideals and turn them into media policy and ideology to which journalists must adhere. This can be a detriment to pluralism and freedom of the press, that is, the dangers of journalists being spokespersons for governments. Even if state leaders raise bold claims about climate change, they must be held to account when it comes to national measures and policy implementation. Thirdly, development journalism should be concerned with climate consequences for "grassroots" people; proactively covering development topics, and giving citizens the chance to voice their concerns in the media (Czepek 2005; Fair 1988; Kunczik 1988; Moemeka 1989; Xu 2009).

Based on the above analysis, we recommend reporting that takes a bottom-up approach to development journalism, as espoused by scholars like Banda (2008), Daorueng (2008), Sharma (2007), Skjerdal (2011) and Xu (2009). This could be dubbed civic or citizen-oriented "advocacy journalism" (Waisbord 2009), focusing on victims of global warming consequences; giving climate change faces and voices. More personalization and visualization of climate stories might help audiences to feel physical and psychological proximity to climate change, and solution-oriented stories may bring hope instead of perpetuating doom and gloom scenarios.

BIBLIOGRAPHY

Aycan, Z. (2002). Leadership and teamwork in developing countries: Challenges and opportunities. *Online Readings in Psychology and Culture, 7*(2). Retrieved from http://scholarworks.gvsu.edu/orpc/vol7/iss2/1/

Banda, F. (2008). Deconstructing media coverage of development. *Media Development, 55*(1), 30–34.

Banda, F. (2013). *Climate change in Africa: A guidebook for journalists.* Paris: UNESCO.

Bond, P. (2012). *Politics of climate justice: Paralysis above, movement below.* Durban: University of KwaZulu-Natal Press.

Brundtland Report. (1987). *Our common future.* Retrieved from http://www.un-documents.net/our-common-future.pdf

Corner, A. (2011). *Hidden heat: Communicating climate change in Uganda: Challenges and opportunities.* Kampala: Panos Eastern Africa.

Czepek, A. (2005). *Pressefreiheit und pluralismus in Sambia.* Demokratie und Entwicklung (Vol. 47). Münster: LIT Verlag.

Daorueng, P. (2008). Is development journalism relevant? In S. Singh & B. Prasad (Eds.), *Media & development—Issues and challenges in the Pacific Islands* (pp. 243–250). Lautoka, Fiji: Fiji Institute of Applied Studies & Pacific Media Centre.

Deane, J. (2009). *Least responsible, most affected, least informed: Public understanding of climate change in Africa.* London: BBC World Service Trust.

EDGAR. (2013). *CO2 time series 1990–2013 per capita for world countries.* EDGAR Emission Database for Global Atmospheric Research, European Commission. Retrieved from http://edgar.jrc.ec.europa.eu/overview.php?v=CO2ts_pc1990-2013

Eide, E., & Ytterstad, A. (2011). The tainted hero: Frames of domestication in Norwegian press representation of the Bali climate summit. *The International Journal of Press/Politics, 16*(1), 50–74.

Fair, J. E. (1988). A meta-research case study of development journalism. *Journalism Quarterly, 65*, 165–170.

George, S. (2009). What is development journalism? *The Guardian*, November 23. Retrieved from https://www.theguardian.com/journalismcompetition/professional-what-is-development-journalism

Germanwatch. (2015). Global climate risk index 2015. Released by Germanwatch. org. Retrieved from http://germanwatch.org/en/download/10333.pdf

Hepworth, N., & Goulden, M. (2008). *Climate change in Uganda: Understanding the implications and appraising the response*. Edinburgh: LTS International.

International Monetary Fund. (2013). World economic outlook database. Released by IMF. Retrieved from http://www.imf.org/external/pubs/ft/weo/2015/01/weodata/index.aspx

Isbrekken, A. (2013). Klimajournalistikk i stormens øye. Journalistisk dekning av klimaendringer på Salomonøyene. Master thesis, University of Oslo, Oslo. Retrieved from https://www.duo.uio.no/handle/10852/38763

ITU. (2013). Global ICT developments. Released by International Telecommunication Union. Retrieved from http://www.itu.int/en/ITU-D/Statistics/Pages/stat/default.aspx

Kunczik, M. (1988). *Concepts of journalism: North and South*. Bonn: Friedrich-Ebert-Stiftung.

Lee, T. M., Markowitz, E. M., Howe, P. D., Ko, C.-Y., & Leiserowitz, A. A. (2015). Predictors of public climate change awareness and risk perception around the world. *Nature Climate Change, 5*, 1014–1020.

McQuail, D. (2003). *Media accountability and freedom of publication*. London: Oxford University Press.

Moemeka, D. A. (1989). Perspectives on development communication. *Africa Media Review, 3*(3), 1–24.

Nassanga, G. L. (2013). African media and the global climate change discourse: Implications for sustainable development. *Open Science Repository Communication and Journalism*, 1–17. doi:10.7392/openaccess.23050451.

ND-GAIN. (2013). The Notre Dame global adaptation index. Retrieved from http://index.gain.org/matrix

Ogan, C. L. (1980, August). Development journalism/communication: The status of the concept. *Paper presented at the annual meeting of the Association for Education in Journalism*, Boston, MA. Retrieved from http://eric.ed.gov/?id=ED194898

Okello, O. F. (2011). Media coverage of climate change in East Africa: Importance, challenges, opportunities and needs. Presentation. Retrieved from http://ssc.bibalex.org/viewer/detail.jsf?lid=D6BE18F19639634228114B8FDAB29992

Page, E. (1999). Intergenerational justice and climate change. *Political Studies, 47*(1), 53–66.

Painter, J. (2013). *Climate change in the media: Reporting risk and uncertainty*. Oxford, UK: I.B. Tauris and Reuters Institute for the Study of Journalism.

Pantsios, A. (2014, October 30). 10 countries facing extreme climate risk. Released by EcoWatch Transforming Green news website. Retrieved from http://ecowatch.com/2014/10/30/extreme-climate-change-risk/

Reporters Without Borders. (2015). *World press freedom index.* Released by Reporters Without Borders. Retrieved from http://index.rsf.org/#!/index-details

Rhaman, M. (2010). Bangladesh: A metaphor for the world. In E. Eide, R. Kunelius, & V. Kumpu (Eds.), *Global climate, local journalism: A transnational study of how media make sense of climate change* (pp. 67–81). Freiburg: Projekt Verlag.

Schramm, W. (1964). *Mass media and national development: The role of information in the developing countries.* Stanford: Stanford University Press.

Shanahan, M. (2009). Time to adapt? Media coverage of climate change in non-industrialised countries. In T. Boyce & J. Lewis (Eds.), *Climate change and the media* (pp. 145–157). New York: Peter Lang.

Sharma, D. C. (2007). *Development journalism: An introduction.* Manila: Asian Center for Journalism.

Shaw, C. (2014, February 6). *Book review: Climate change in the media: Reporting risk and uncertainty by James Painter.* Retrieved from http://blogs.lse.ac.uk/lsereviewofbooks/2014/02/06/book-review-climate-change-in-the-media-reporting-risk-and-uncertainty/

Skjerdal, T. (2011). Development journalism revived: The case of Ethiopia. *African Journalism Studies, 32*(2), 58–74.

Statistics Norway. (2013). Statistisk sentralbyrå. Retrieved from http://www.ssb.no/en/

Synovate & Deutsche Welle. (2007). Climate change concerns remain high across the globe. Retrieved from http://www.dw-gmf.de/start/5106.php

Toulmin, C. (2009). *Climate change in Africa.* London, UK: Zed Books.

UN Development Programme. (2013). *Human development reports by United Nations Development Programme.* Retrieved from https://assignmentbro.com/blog/human-development-index-and-its-components

Waisbord, S. (2009). Advocacy journalism in a global context. In K. Wahl-Jorgensen & T. Hanitzsch (Eds.), *Handbook of journalism studies* (pp. 371–385). New York: Routledge.

Wiggins, M., & Liu, L. (2010). Climate change profile Uganda. Tearfund. Retrieved from http://www.tearfund.org

Wimmer, J. & Wolf, S. (2005). Development journalism out of date? *Munich Contributions to Communication, 3,* 1–12. Retrieved from http://epub.ub.uni-muenchen.de/647/1/mbk_3.pdf

Xu, X. (2009). Development journalism. In K. Wahl-Jorgensen & T. Hanitzsch (Eds.), *Handbook of journalism studies* (pp. 357–370). New York: Routledge.

Zamith, R., Pinto, J., & Villar, M. E. (2012). Constructing climate change in the Americas: An analysis of news coverage in the U.S. and South American newspapers. *Science Communication, 35*(3), 334–357.

CHAPTER 11

Good Practices in Climate Science Journalism

Elisabeth Eide and Oliver Hahn

The best reporters look for options.

Chris Field, IPCC WGII author[1]

An analysis of "good practices" in journalism covering the IPCC AR5 needs to be normative by the sheer nature of the effort. This chapter has looked for journalistic articles that satisfy at least some criteria that are crucial for bridging gaps in science communication, which is oftentimes deemed as elitist and distanced from people's everyday experiences (Beck 2010).

Climate journalism has been subject to criticism for being too focused on the political game (Naper 2014), too biased (Boykoff & Boykoff 2004) or disaster oriented (Painter 2013), at times promoting a debilitating alarmism

---

[1] Chris Field was interviewed by Elisbeth Eide in Copenhagen on August 20, 2014 and again in Paris on July 8, 2015 by Elisbeth Eide and Risto Kunelius.

E. Eide (✉)
Department of Journalism and Media Studies, Oslo and Akershus University College of Applied Sciences, Oslo, Norway

O. Hahn
Centre for Media and Communication, University of Passau, Passau, Germany

© The Author(s) 2017   235
R. Kunelius et al. (eds.), *Media and Global Climate Knowledge*,
DOI 10.1057/978-1-137-52321-1_11

(Risbey 2008). One may easily fall into a trap and believe that if a "best" or "good practices" formula is found (at least some), media will improve their climate change coverage and, thus, the public will follow suit and opt for more climate friendly practices. However, the influence of media is limited, with many other factors playing a role in the decisions made by politicians and ordinary people (Austgulen 2012). Second, even if people know more in-depth what is best for the planet and its future, it is far from certain they will act accordingly, given this may entail the loss of a privileged, carbon-driven lifestyle, as Leon Festinger suggested in his outline of human "cognitive dissonance" (1957). Or, as Anabela Carvalho iterates,

> Divided between the belief in forecasts of devastating impacts of climate change and the faith in reassuring reports on the ability of politicians and scientists to solve the problem, the common citizen feels confused and powerless. But set against short-term needs and wants and embedded in a deeply material and mobile civilization, that same citizen prefers a strategy of denial. (Carvalho 2005: p. 15)

Ten years later, the forecasts are no less devastating, but the reports perhaps less reassuring since global GHG emissions are still on the rise. The mental urge to resort to denial or negligence may be just as strong among lay people, which poses a formidable challenge to journalism.

HOW TO DEFINE "GOOD PRACTICES"?

The concept of "good practices" is of interest to journalists and journalism researchers alike. Covering a global crisis that will certainly not go away presents a major challenge for journalism, not least when it comes into conflict with daily routines and practices that are, most of the time, confined to operate within the borders of the nation. Climate change in a fundamental way challenges journalism's general conventions and doxa (Benson and Neveu 2005; Lakoff 2010; Eide and Kunelius 2012), mainstream news being concerned with the abrupt and episodic, overshadowing slower processes such as climate change that threaten livelihoods globally.

Many researchers have addressed the issue of journalism on climate change. A particular challenge is that while climate science in all its complexity of certainties and uncertainties may be hard to communicate to lay people, climate skeptics or denialists can exploit this dilemma by adhering to populist, anti-elite rhetoric (Myers 2014). Painter (2013) demonstrates how print media in six countries increased the number of articles containing skeptical voices between 2007 and 2009–2010 (58–59), and further-more that climate deniers "had been marginal to the scientific debate, but somehow they continued to find a place on the airwaves" (Painter 2013: p. 111). Explanations for this tendency include the existence of strong lobby groups that support skeptics and a penchant for conflict and polar-ization among the media. Consequently, the findings emphasize the need for increasing climate consensus, as well as the precautionary principle when it comes to climate change, in order to reach a wider audience of lay people. Secko et al. (2013), who differentiate between "traditional" and "non-traditional" models of science communication, state that the traditional top-down "science literacy model" is "at the heart of some recurring critiques leveled at science journalism" (p. 75), a critique which also addresses the need for information that reaches out to people at the grassroots level.

Generally, journalistic ideals include aiming for a democratic pub-lic sphere, being (auto)critical, taking sources seriously without being exploited, being knowledgeable, humble and, not least, reporting the situation "on the ground" (Eide 2011). The latter concern relates to what James Hansen and others call "connecting the dots" (Hansen 2014); one of the key connections being between science and people's everyday experiences. Investigative reporting on the politics and effects of climate change linked to the communication of climate science are criteria to be considered, as are accuracy and truthful reporting on the existing uncer-tainty of science communication. Climate researchers are careful when communicating, for example, about climate change processes linked to extreme weather, although they conclude with increased frequency (IPCC 2012, 2014). This interpretational gap between lay people, apprehend-ing a change in local weather conditions, and scientists, who take care to explain the complexity of climate change, is one of the main challenges for producing "good" climate journalism.

One of journalism's central ideals is to inform the public so they can make wiser decisions about their own lives. In this chapter, we have looked for reporting which tries to connect IPCC research to people's everyday lives and provide some space for views from the ground. We also identify reporting which manages to explain the research in a balanced and engag-ing way. Finally, in the study, we consider news items that use creative techniques to challenge politicians and their responsibilities, as well as well-written opinion pieces with a certain amount of explanatory content. Our sample stems from a call to partners in the *MediaClimate* network (from 22 countries) who were asked to nominate stories from the AR5 coverage that they thought would qualify as "best" or "good" practices.

In an editorial for *Public Understanding of Science*, Martin Bauer (2009: p. 378) mentions three different attitudes toward "common sense". We consider them relevant for journalism since enhancing "science literacy" must take common sense (in itself a "contested territory") into account. These three attitudes are:

– Debunking of traditional beliefs, including the identification of "deficiencies of public opinion and common sense";
– Intervention: "We need reports that take stock of the deployed resources, evaluate this mobilization effort, and measure and assess the public presence of science across the world and in very different contexts" (Bauer 2009: p. 379). Intervention is seen as a way to "improve science literacy, mobilize favorable attitudes in support of science and new technology, [...] and to intensify the public's engagement with science in general and for the greater good of society";
– Common sense as a "*resource* of inspiration, oversight and legitimacy", enhancing local knowledge, traditions and social capital. "It is risky to dismiss the non-scientific mentality a priori as ignorance in the face of uncertainty". Here people and common sense are viewed as representative of the "real impacts" of science and scientific evidence (Bauer 2009: pp. 379–380).

Bauer's points on common sense and public engagement further underline the need to involve people and—without disregarding critical assessment—take people's livelihoods and local experiences into account when working with science journalism. Not all journalists can meet this whole row of criteria; however, in our research we discovered clear examples of journalists in print media trying to integrate them into their general engagement with climate change, and in covering the IPCC reports in particular. In this chapter, we have also integrated viewpoints from leading IPCC authors, interviewed prior to and during UNESCO's "Our Common Future under Climate Change" conference in Paris in July 2015.[2] A recent Reuter's Institute study on quality journalism emphasizes democracy as an important precondition for quality, while adding several other elements including relevance, "context, background, interpretation or analysis. It should [also] be surprising, interesting, and smart"

---

[2] The interviews were conducted by Elisbeth Eide and Risto Kunelius.

(Vehkoo 2010). While no single journalistic item could fulfill all these expectations or criteria, in this chapter we identify items containing at least some. Our material is clearly limited to findings in the media selected in the participating countries. Previous studies have found that the reportage genre holds potential for including marginalized groups and for communicating a sense of humanity by exemplifying and personalizing experiences linked to complex phenomena (Carey 1989; Hartsock 2009).[3]

From the Arctic and Further South

This chapter presents stories that demonstrate "good practices" from newspapers published in the "global South", such as Bangladesh, Indonesia and Uganda, as well as items from the "global North", with some of the latter addressing problems and challenges in developing countries. One obvious obstacle has been the low coverage of the AR5 reports, particularly in the "global South", and our exclusive focus on a limited sample of *newspaper* coverage (see Chap 1, especially Sidebar 1.1, for details on sample and methodology). Although by no means representative of climate journalism in the member countries nor of "best practices" in science journalism, the empirical material presents a variety of coverage that speaks to some aspects of "good practices". Stories vary in content, from reporters who ventured to areas illustrative of climate change, some to the extreme North and others to island nations, like Maldives, considered vulnerable to sea rises caused by global warming, to reporting "from home". Some items are comments, some straightforward news, and others labeled as feature stories. Earlier findings indicate the feature story genre holds the most potential for explaining climate change to lay people and for describing the ways in which ordinary people may experience climate change happening now (Eide and Kunelius 2012).

Among the Ice and Snow

Journalism that mainly communicates scientific knowledge is represented by three reports from two Norwegian newspapers: *Aftenposten* (2) and *Dagsavisen* (1). Each concentrates on the consequences of climate change in the Arctic region. One story points to the fact that many IPCC results stem from studies in the Arctic, where phenomena such as melting ice

---

[3] This is exemplified by new online journalism platforms, such as CODA story, whose intention it is to "stay with a story": http://www.codastory.com/about

caps are occurring.[4] In another story, scientists recount earlier ice ages and describe abrupt changes expected for the Arctic ice.[5] A third item presents a report indicating that animal life, not least reindeer in Spitsbergen, may be seriously affected by global warming.[6] Scientific uncertainty is perhaps best illustrated in the article on climate change in Greenland:

> This summer has also seen more melting than normal for Greenland, but far less than during the record year in 2012. [...] climate models show that the meltdown of ice in Greenland will take several thousand years. Imagine if this is wrong, and that it happens much faster?

The story includes beautiful, almost sculptural photos from Greenland, several accompanied by alarming captions. For example: "Large [ice] melting on Greenland happens every 150 years. The recent large melting has occurred almost every year".[7] Debates among scientists are referred to in the form of several questions that scientists are struggling to answer. Thus, the reporter demonstrates that the science is not all "settled". The main source is a statement, by a Norwegian lead author of AR5 WGI, predicting that we "may have ice-free oceans in the Arctic in the last half of this century" and highlighted with graphics. All three stories demonstrate the importance of global co-operation between scientists who, unlike politicians, manage to organize themselves for transnational research endeavors.

While it was foreseeable that Norwegian journalists would focus on the Arctic region due to national interests and activities there, it was equally probable that journalists and commentators in Bangladesh would focus on climate adaptation, being a poor and vulnerable country with little historic responsibility for climate change. In a comment for *The Daily Star*, Saleemul Huq, a Bengali climate scientist (based in the UK), relates particularly to the AR5 Synthesis Report (November 2014). The report predicts that a four-degree Celsius global temperature rise will "lead to globally catastrophic levels of adverse climate impacts, to which not just Bangladesh but even the rich world will find hard to adapt". He also mentions the need to phase out greenhouse gas by the turn of the century. Huq adds that while Bangladesh is in need of adaptation measures, his country must also focus more on mitigation. He explains why in scientific terms: "...while adaptation is necessary and must be done in the short to

[4] Rapp, O. M. (2013). Everyone wants to do research in Spitsbergen. *Aftenposten*, October 8.
[5] Mathismoen, O. (2014). What is really happening in Greenland? *Aftenposten*, November 16.
[6] Prestegård, S. (2014). Warmer weather threatens animal life. *Dagsavisen*, November 20.
[7] Mathismoen, O. (2014). What is really happening in Greenland? *Aftenposten*, November 16.

medium term, there is a limit to what can be achieved by adaptation, as it will never bring down the adverse impacts to zero".[8] While his critique is directed at the one-sided focus of national politicians on adaptation, his main focus is on the responsibility of Bangladeshi leaders to pressure rich nations to increase their mitigation efforts. His conclusion is that Bangladesh must "focus more on adaptation actions domestically". This is in line with Mike Hulme's "climate pragmatism" (2014: p. 122) where one of the main points is to increase the resilience of a society to ongoing changes.

In a strict sense, this is not journalism, but rather an outwardly focused opinionated article. However because the article is written by a Bengali expert, it exemplifies how this Bangladeshi newspaper feels the need to communicate climate science by way of domestication, finding a voice who knows realities within the country. It is published in the country's elite, English language newspaper, and written in a manner which communicates well to this elite, underlining political and scientific aspects and the need for enhanced efforts toward the Paris summit (COP21).

In another opinion piece, published in Germany's leftist paper *Die Tageszeitung (taz)*, Bernhard Pötter evaluates the concession of the "dirty three", the USA, the EU and China, to take responsibility and give way for climate negotiations.[9] Yet the goals and numbers presented by the big three are explained to readers and relativized as only "minimum bids in the poker game until Paris". Interestingly, this comment emphasizes that "the fight against carbon does not only take place during climate conferences, but also on the streets", and that "economic pressure" is needed: "those who will lead in the 'clean energy' market will be one step ahead in the 21st century". This op-ed provides illustrative transparency to lay readers by "translating" political promises expressed through easily understandable numbers that are not oversimplified as well as "down-to-earth" measures and consequences. Here, the emphasis is on popular action, and on meeting future energy demands, leaning toward a grassroots oriented discourse that contrasts the *view from the ground* with the power players heading for the negotiation table, much in line with Banda's criteria for development journalism (2008).

[8] Huq, S. (2014). Why climate change mitigation matters for Bangladesh. *Daily Star*, November 14. Retrieved from http://www.thedailystar.net//why-climate-change-mitigation-matters-for-bangladesh-49229.

[9] Pötter, B. (2014). Does the cleaning begin? *Die Tageszeitung*, September 15.

COFFEE: CROPS, FARMERS AND CONSUMERS

Coffee is part of many people's daily lives, and considered a necessity. Interestingly, three newspapers, one from Uganda (*Daily Monitor*), one from Australia (*Sydney Morning Herald*) and one from the UK (*The Guardian*), all address the dangers global warming poses for coffee growers.

"Temperature rise threatens coffee production" is a news item from Uganda with some opinion features that lean toward a business focus[10]. The IPCC is not mentioned directly, on the contrary, the text refers to a "National Water Resources Assessment Report" which predicts that "rising temperatures in the country could reduce by 85 percent the available land used for growing Robusta coffee". The headline captures the main message and the story is illustrated by two hands picking coffee berries. A caption echoes the main message by calling coffee "one of Uganda's most lucrative products". Bullet points listed under a separate headline highlight "Climate change effects", referring to an expert from the National Agricultural Research Organization (NARO) who says that "climate change has been visible in the country although not much attention has been given to it". The section also cites a former Minister of Agriculture who asks the government to allow scientists to give farmers improved seeds to help them avoid "crop declines that will have a negative impact on rural incomes". Here, the newspaper highlights consequences for the coffee growers; however, no farmers are quoted in the article. The scientific statements are echoed in the main text, where an officer at Uganda Coffee Development Authority states that "[the] predictions have been there for the last 10 years. It is not new". The article ends, however, by showing numbers indicating that coffee exports from Uganda continue to increase.

The article promotes a *climate change vs. national interest-discourse*, by focusing on the main cash crop in Uganda, and by only quoting national sources, be they experts, business people or politicians. It is very much *adaptation*-oriented, in the sense that new crops, more adaptable to climate change, must be found to maintain farmer livelihoods and the economy of the nation. While the sources quoted are all elite, the picture of coffee picking hands can be viewed as a metonym, standing in for Ugandan farmers absent as voices in the story yet implicitly present in references to their livelihoods. As an article on climate change published after the launch of the

---

[10]Tajuba, P. (2014). Temperature rise threatens coffee production. *Daily Monitor*, November 11. Retrieved from http://www.monitor.co.ug/Business/Prosper/Tempreture-rise-threatens-coffee-production/-/688616/2517510/-/3p8a8q/-/index.html

AR5 Synthesis Report, it relates general predictions from a global scientific body (the IPCC) to one of the main sources of income for Ugandans.

From another corner of the world, and linked to the publication of the AR5 (WGII), *The Guardian* published the story "How climate change will brew a bad-tasting, expensive cup of coffee".[11] The lead continues: "First it was glaciers, now your coffee faces climate threat". This differs from the Ugandan story in terms of focus: there on *production*, here on *consumption* ("your coffee"). The article opens by addressing coffee consumers with a hint of irony:

> Rich western urbanites expecting to dodge the impacts of climate change should prepare for a jolt: global warming is leading to bad, expensive coffee. Almost 2 billion cups of coffee perk up drinkers every day, but a perfect storm of rising heat, extreme weather and ferocious pests mean the highland bean is running out of cool mountainsides on which it flourishes.[12]

This angle resonates with an Australian story, which also expresses irony toward urban coffee-drinking elites:

> The latte sippers of the inner city may not be the communities most threatened by climate change, but the latest UN scientific assessment of global warming's impacts suggests they will not go untouched. That is because the Intergovernmental Panel on Climate Change projects significant disruption to coffee production as the planet warms.[13]

A "world champion barista" who is interviewed says that "the real pressure will be felt by the millions of small landholder producers of coffee in Africa, Asia and Latin America". Another source focuses on how climate change will open up new areas for coffee growing, substituting for those that may have to stop. The story's main tone is pessimistic, however, anticipating that coffee taste and quality may decline while prices go skyrocketing upwards.

As *The Guardian* story develops, it refers directly to the WGII report and its predictions of a decline in coffee production due to temperature

---

[11] Carrington, D. (2014). How climate change will brew a bad-tasting, expensive cup of coffee. *The Guardian*, March 28. Retrieved from http://www.theguardian.com/environment/2014/mar/28/climate-change-bad-expensive-coffee-ipcc.

[12] Carrington, D. (2014). How climate change will brew a bad-tasting, expensive cup of coffee. *The Guardian*, March 28. Retrieved from http://www.theguardian.com/environment/2014/mar/28/climate-change-bad-expensive-coffee-ipcc

[13] Arup, T., & Wood, S. (2014). Temperature rise could hit coffee drinkers in hip pocket. *Sydney Morning Herald*, March 31.

rise, pests and more extreme weather, all negative outcomes for crops. Three sources are quoted in this text, the WGII report, the executive director of World Coffee Research, and a head of operations representing the Intergovernmental International Coffee Organization, with the latter two expressing similar concerns. *The Guardian* story also stresses that in Uganda: "… adaptation by shifting plantation up hillsides will be impossible: they will simply reach the top and run out of land". However, it ends on a note of optimism with a scientist referring to the genetic variety of coffee out of which some may be "crossed and put into field trials" to develop what one researcher calls "super races of coffee"; adding, however, that experts are running late and should have begun experimenting ten years ago. The story ends with the punch line: "But I wonder if coffee growers will be able to withstand climate change for another 10 years". Thus, a text that began with urbanite consumers ends with vulnerable growers and an attempt to bridge the gap between consumers and producers.

Similar to the report from Uganda, *The Guardian* story does not quote or mention coffee farmers. The two also chose related visualizations. The first *Guardian* image depicts a hand holding and spilling coffee beans. The caption describes this as "a worker checking roasted coffee beans at a roasting plant in Cavite, Philippines".[14] Another similarity is that both stories represent a discourse of *technological optimism*. Research on adaptation will ease the problems; agro-engineering may do the job, although as stated in *The Guardian* article, many countries will lose their ability to produce coffee. The social impact on people in Central America, Laos, Peru, Burundi and Rwanda, who are among the losers, is mentioned:

> At least 1.4 million people in Guatemala, El Salvador, Honduras and Nicaragua depend on coffee production for their livelihoods. When coffee's susceptibility to changes in climate has caused crises in the last few decades, a quarter of all households have been forced to migrate.

Thus, all three stories pay some attention to the threatened livelihoods of coffee farmers; however, they are produced without venturing out to meet the human beings behind these "livelihoods". The articles communicate the science and are relatively faithful to elite expertise, be they scientists or top executives, but not representative of the "view from the ground".

---

[14] The two other photos depict fields in Brazil (the world's biggest coffee producer) and crops damaged by fungus in Guatemala.

BIOGAS HOPE IN UGANDA

One of journalism's ideals is to provide the public with guidance, but far too often science journalism is about catastrophe, simplifying and exaggerating to draw attention to a story. As Swyngedouw (2010: p. 24) iterates, "an ecologically and climatologically different future is only captured in its negativity; a pure negativity without promises of redemption, without a positive injunction that 'transcends'/sublimates negativity and without proper subject". Swyngedouw's emphasis is on the political more than on media representation; only rarely does one find climate change stories communicating hope for the future.

One of the rare hope oriented items in this sample is a story on biogas from Uganda: "Biogas: A school is using a latrine to cut its energy bill".[15] The story relates to the IPCC summary report, but starts in a suburb in Uganda's capital city, Kampala, where the community has found "a way of minimizing the bills they pay for energy while reducing their carbon footprint, all with the help of a latrine". The reporter explains that this simple technology "[...] has the potential of turning organic wastes into sources of clean energy that can save the country's forest cover and also minimize air and water pollution". The sources quoted include the school's head teacher, Musa Busa Mwima, a biogas chief engineer from an NGO, another NGO leader and the IPCC report. It also refers to Ugandan President Yoweri Museveni's 2014 speech at the 69th UN General Assembly, where he claimed "climate change is a new modern aggression against the always wronged African continent and its people".

The story strives for a holistic approach, highlighting the multi-dimensional effects of biogas technology: hygiene, it does not smell or attract flies; an energy source for school stoves, thus conserving energy and forests; creates a clean cooking environment by reducing in-house pollution from charcoal and firewood; and hence reduces the community's ecological footprint. The text allocates some paragraphs to explaining the science. It refers to the IPCC report and does not exclude African emissions: "The main activities causing climate change have been identified as greenhouse gas emissions, burning of fossil fuels, extraction of minerals, deforestation and mushrooming urban settlements in Africa among

[15] Ainebyoona, E. (2014). Biogas: A school is using a latrine to cut its energy bill. *Daily Monitor*, November 6. Retrieved from http://www.monitor.co.ug/artsculture/Reviews/ Bio-gas--A-school-is-using-a-latrine-to-cut-its-energy-bill/-/691232/2511926/-/ jj9a3x/-/index.html

others". Furthermore, it mentions the first decade in the new century as the warmest since 1850 and refers to unpredictable rain patterns in Uganda. Finally, the story lists a range of other consequences that might be attributed to climate change: landslides, bursting of riverbanks, melting of snow peaks and the flooding of roads in Kampala.

The article also includes an appeal for climate justice: "Environment activists have been demanding that industrialized economies which are the largest contributors of greenhouse emissions should fund clean energy projects in Africa as a way of mitigating climate change". This feature story represents a *discourse of hope* through simple, people-friendly technological innovation, while also relating to the global situation described by the IPCC and to national politics. An underlying discourse in the latter is the *discourse of climate justice*, attributed to and clearly adopted by Uganda's president. Thus, a confrontation with politicians is downplayed while emphasis is placed on what can be done if more funds are provided by the so-called rich world. To conclude, the reporter returns to the school and its head teacher, praising the latrine for improving children's hygiene. The author addresses the reader directly in a rather pedagogical way by "breaking the fourth wall": "When you factor in all the above, there is no denying that any effort no matter how small it seems, as is the case with the school in Kansanga, is important in the fight against climate change". This "view from the ground" becomes an example of how climate change might be fought for the benefit of local people.

Solar Hope in Texas and Micro-Loans in Bangladesh

The discourse of hope through larger-scale technological innovation is present in a report by the Norwegian *Dagsavisen's* US correspondent.[16] The reporter travelled to Austin, Texas where she met a woman who works at "a giant solar farm" outside the city. The employee proudly explains that Austin will reach its target of using 35 percent solar energy before 2020, maybe even as early as 2016. She adds that "Green energy is now almost as cheap as fossil fuel energy in the oil state of Texas", a fact that may be surprising for many readers. "This is the future. Yes, we have lots of oil and gas. But the sun is free and always here", says the farm's man-

---

[16] Skjeseth, H. T. (2014). Solar energy advancing. UN climate panel: Climate friendly energy must be tripled by 2050. *Dagsavisen*, April 28.

ager, who also boasts of having sheep brought in to diminish growth that can hamper the effects of the solar panels. The story proceeds to highlight the investment potential, quoting a Norwegian expert who endorses solar energy. It ends on a more pessimistic note, referring to the fact that 20,000 new oil and gas wells are drilled every year in Texas. The story is illustrated with graphics explaining solar panel technology and a large photo of an enthusiastic staff member.

A story from *The Daily Star* in Bangladesh starts on a pessimistic note, envisaging negative outcomes for the country's Micro Finance Institutions (MFIs), as more farmers fall into debt because of crop failures caused by climate change. Bangladesh is historically famous for its micro-credit schemes, and this story uniquely relates them to IPCC climate science. Fazle Rabbi Sadeque Ahmed, the head of a community climate change project, is interviewed and does not rule out new opportunities. As the article concludes: "The MFIs are already involved in a limited scale with adaptation activities such as introduction of new, resistant and improved varieties, adaptive cropping pattern, adaptive and resilient housing and alternate income generating activities".[17] This article advocates for improving the usage of an important and successful element, which has empowered many poor citizens in Bangladesh, especially women, and suggests a way in which micro-credit can serve as a tool to enhance adaptation measures in the society at large.

THE VIEW FROM JUST ABOVE THE WATER

In a report filed the day before the release of WGI, Damien Carrington travelled to the Maldives, another lowland nation highly sensitive to climate change.[18] The text begins dramatically:

> On Friday 27 September, the low-lying island nation of the Maldives will be given the date of its extinction; notice of a death by drowning. It will come in the form of a prediction for future sea-level rise in a landmark report on global warming by the world's climate scientists.

[17] Micro loans can offset effects of climate change: Analyst. (2014). *Daily Star*, November 16. Retrieved from http://www.thedailystar.net/micro-loans-can-offset-effects-of-climate-change-analyst-50527

[18] Carrington, D. (2013). The Maldives, a fledgling democracy at the vanguard of climate change. *The Guardian*, September 26. Retrieved from http://www.theguardian.com/environment/2013/sep/26/maldives-democracy-climate-change-ipcc

Carrington finds himself in Male, the capital, illustrated by an aerial photo-graph depicting a flat, tiny and overcrowded island. He writes of political turbulence, adding that climate change "was as invisible during the country's election campaign as the carbon dioxide that drives it". This observation may have inspired the reporter to contact former Minister of Environment Mohamed Aslam; who actively promoted the issue before COP15 in Copenhagen by staging the first ever underwater government conference, photos of which became widespread in media across the world.[19] The "view from the ground" is summed up by a taxi driver, Mohamed Arzan: "If I woke up one day and the islands were going underwater, I'd just move". Apart from this voice, Carrington's sources are all elites, including four national scientists/experts and one British engineer. They each warn against the political negligence of climate change and spell out the variety of con-sequences already clearly evident in this island state. Among them the dete-rioration of coral reefs; an increase in the number of severe weather events; coastal erosion; rising ocean temperatures that directly affect the tuna industry (second only to tourism in terms of national income); extended, intensified dry seasons; and problems with drinking water.

The article is clearly IPCC-relevant, as the reporter demonstrates an awareness of the main conclusions in the AR5, relating them to the increasingly serious problems in this vulnerable island state. An ironic tone emerges in several paragraphs that characterize a recent election campaign on the island nation that was free of climate issues. Toward the end of the text emphasis is placed on an uncertain future:

> An entire new island, called Hulhumalé, has been created from sand dredged from the lagoons. When the second phase is finished, the 420 hectares will house 160,000 people, a third of the entire population. By Maldivian stan-dards, the new island is lofty, at 2m high. "It is the safest island in the Maldives, from sea level at least", says Ahmed Varish, at the Hulhumalé Housing Development Corporation, standing outside the large social housing blocks.

In the final paragraph Carrington adds his own commentary, stating that the new IPCC report will judge how safe this really is. He ends with a quote from former Minister of Environment Mohamed Aslam: "Whatever

---

[19] See for example: Maldives government highlights the impact of climate change... by meeting underwater. (2009). *Daily Mail*, October 20. Retrieved from http://www.daily-mail.co.uk/news/article-1221021/Maldives-underwater-cabinet-meeting-held-highlight-impact-climate-change.html.

happens to us will happen to the rest of the world—it is just a matter of time". Combined with the first paragraph of the story, it is tempting to conclude that this reportage resorts to an *alarmist discourse*: "extinction", "death by drowning" and "existential threat". However, the dominant use of local scientists, who present their own findings, opts for balance and recent experiences from other low-lying islands such as Vanuatu, underline the *alarming* and precarious situation for these countries.[20] The distinction between *alarming* and *alarmist* may at times be apprehended as thin, but remains essential (Risbey 2008) since the latter is related to a "Cry wolf-syndrome" and to disaster-discourses, while the former describes scientifically established threats.

Addressing Uncertainty and Other Difficult Issues

Triggered by the global "warming pause", some have cast new doubts on the established threats communicated in the IPCC's scientific results. The "pause" has to do with the relative slowing down of the earth's surface temperature during the last decade. The French newspaper *Le Monde* dedicates a full page to WGI, illustrating the findings with a global map that displays the variety of predicted consequences of continued global warming. The text is divided into smaller sections: temperatures, glaciers, sea level, extreme events and geo-engineering. *Le Monde* treats the "pause" issue in a sub-article below its main presentation of WGI findings. [21]

This story explains the variable and unstable character of temperature rise, pointing to a slowdown that occurred in 1998 (which was also the year when El Niño caused extreme temperatures), while the warming rose more in previous decades. The text also mentions that certain scientific models may be inadequate. The linguistic ambiguity in the headline demonstrates this complexity: "La vraie-fausse 'pause' du réchauffement" ("The true-false of the warming 'pause'"). The sub-article, however, conveys a clear message: the WGI report places "additional pressure on the international negotiations", and warns that the economic crisis has overshadowed cli-

[20] See for example: Cyclone devastates South Pacific Islands of Vanuatu. (2015). *BBC News*, March 14. Retrieved from http://www.bbc.com/news/world-asia-31883712 and Cyclone relief in Vanuatu continues over Easter despite difficult conditions. (2015). *TVNZ One News*, April 5. Retrieved from https://www.tvnz.co.nz/one-news/new-zealand/cyclone-relief-in-vanuatu-continues-over-easter-despite-difficult-conditions-6277919.html
[21] Caramel, L. (2013). A possible heightening of 4.8 from now until the end of the century. *Le Monde*, September 28.

mate urgency. Most importantly, it tackles head on one of the anticipated controversies linked to the launch of AR5 WGI (see also Chap. 2).

Complexity is also addressed in an editorial comment on the phenomenon of *clouds* published in the Indonesian newspaper *Kompas*.[22] This philosophical-historical opinion piece grapples with the relation between human beings, weather and climate in general and relations to clouds in particular. A question is raised:

> Climate scientists observed a series of confusing changes that are taking place on land, sea and air. Then, where should they focus in order to be able to uncover the secrets of these changes? The answer is the clouds. (…) Do heat absorption and radiation back into the atmosphere by clouds contribute to the earth's temperature increase? Alternatively, do heat waves reflect back to the sun so that the Earth's temperature decreases? What is more powerful, the first or the second?[23]

The writer, well-known climate journalist Brigitta Isworo, refers to an IPCC author who answers by saying that the effects of low clouds cannot be ascertained and calls low clouds the "wild card". In other articles, Isworo endorses the IPCC findings and the dangers of climate science; however, here she ventures out to question a particular issue to which the climate scientists have no definite answer. Thus, she successfully integrates a *discourse of uncertainty*, highly valued by scientists but often neglected by journalists who opt for simplicity.

CONSUMER RESPONSIBILITY

A reoccurring discourse in climate change coverage focuses on citizen and consumer responsibility. Consumer and citizen do not mean the same thing, and critical journalism studies have primarily focused on the increased commercialization of journalism, where one visible result is treating the reader/spectator largely as a consumer (Bech-Karlsen 1996; Heinonen and Luostarinen 2008). When it comes to climate change, the two may more easily be seen as intertwined, as the duty of many citizens would be to consume less, or at least in a more eco-friendly manner. A Swedish newspaper treats one consumer-related issue in an article headlined "The

---

[22] Isworo, B. (2014). Uncovering the veil of clouds. *Kompas*, October 11.
[23] Translated from Bahasa Indonesia by Jamal bin Ishak.

difficult dilemma of travelling".[24] This dilemma is described well, focusing, on one hand, on the IPCC WGII report that the current state of affairs may lead to four rather than two degrees of temperature rise and, on the other, a UN prognosis suggesting that global tourism will double toward the year 2030. The text published in the newspaper's travel section, supposedly one of the most consumer-oriented parts, contains a hint of irony about Swedes being concerned with sorting garbage and recirculation, but much less with their travel footprint. An expert states that citizens, otherwise considered climate conscious, leave their consciousness at home during holiday travels. Experts also stress that tourism in previously snow clad areas, such as the Alps, may be badly affected by climate change.

The experts agree that the lack of political steering and guidance toward more environment friendly choices leaves climate responsibility to the individual. Nevertheless, they do suggest some broader amendments such as hotels serving locally sourced food and people staying longer when travelling long distances. They also welcome the development of airplanes that consume less energy. The one-page article is illustrated with an image that demonstrates the magnitude of airplanes over Europe "at lunchtime last Wednesday", and concludes by depicting four future scenarios for tourism in the year 2023; plotted on a scale between freedom without restrictions and a tough regulatory regime. The deeper tone of the story relates to the conflict of ideals versus practices. This resonates with Leon Festinger's theory of cognitive dissonance (1957), which entails a lack of correspondence between convictions and practices, between what one holds up as a moral standard and what one practices in everyday life, driven by self-interest or an urge for the so-called good life.

THE SCIENCE PERSPECTIVE

When asked for their assessment of good climate (science) journalism, the group of IPCC authors interviewed for this study all mentioned the need to cultivate good relations between the science community and journalists. Valérie Masson-Delmotte (WGI) has experienced journalists "who push me to speak outside my area of competence". She associates good practice with reporters who have read the IPCC reports or, at least, the summaries for policy makers, and are thus able to ask "what is new here?" (Interview, July 9, 2015). She also emphasizes the need for the integration of climate

---

[24]Wickström, A. (2014). The difficult dilemma of travelling. *Svenska Dagbladet*, April 6.

science into journalism training. Norwegian WGI author Eystein Jansen has also experienced some journalists, particularly TV reporters, arriving "with a set angle and just want me to fit in. They have already made up their minds". He finds it "easy to communicate with journalists who [...] have followed the topic for some time", but admits that there are only "a handful in Norway" (Interview, May 15, 2015).

The co-chair of WGII, Chris Field says that "Climate journalism should be more solution oriented", adding that this "...opportunities angle would opt for new kinds of coverage, combined with local angles. And the solution story is not separate from the urgency story" (Interview, July 8, 2015). Thomas Stocker, the co-chair of WGI, addresses the ongoing downsizing of newsrooms and the *structure* of the journalistic field, advising the CEOs of media outlets to "re-employ science journalists" and journalists to in turn establish good relations with the scientific community. Alluding to the linguistic gap between high-level technical expertise and a profession that addresses the public, he also stresses that it is important to find ways to "Listen to colleagues who have difficulties expressing themselves" (Interview, July 7, 2015).

Saleemul Huq of WGII recommends more emphasis on feature stories which offer room for communicating various aspects of the climate science. He states that the level of climate awareness in Bangladesh is high, but that editors frequently outsource climate stories to external writers instead of investing in improvements to in-house skills (Interview, July 8, 2015). Some scientists worry that the same climate story is being told and retold. Karen O'Brien's (WGII) advises to try "to find a different perspective, and help readers connect the dots", mentioning the drought in California as an illustrative example. (Interview, July 9, 2015). The scientists' views may be seen as corresponding with some of the stories highlighted in this chapter. Some stories are clearly produced by skilled and specialist reporters (*Kompas, The Guardian*), some are people-oriented and focused on local solutions, trying to "tell new stories" or "connect the dots". Yet, the scientists' fear that climate change is not prioritized enough in editorial rooms remains justified.

CONCLUSIONS AND PERSPECTIVES

To conclude, we shall consider how the stories discussed above correspond to Bauer's (2009) criteria for enhancing the public understanding of science: debunking of traditional beliefs, intervention and common-sense-oriented science communication. The articles in *Le Monde* (France) and

Kompas (Indonesia) may be understood as "debunking popular beliefs" among lay people as well as journalists, by complicating the science and thereby promoting science literacy. The stories on coffee production may be read as assessments of "the public presence of science across the world and in very different contexts" (Bauer 2009: p. 379). When taken together, these three articles represent a prominent example of "contrapuntal reading" (Said 1994) since they assess the threat climate change poses for coffee production from both the point of view of producers and consumers. Several items address transnational issues, such as the reports from the Arctic and *The Guardian* story on the Maldives. Local, "common-sense" wisdom, however, is poorly represented which was to be expected due to the nature of our sample (Chaps. 1 and 4). Some other explanations for the lack of views from the ground may be poor editorial resources and deficits in journalism education. Several stories address what Mike Hulme (2014) calls *climate pragmatism*. To do so they speak to people's needs by making local communities more resilient to climate change threats (*The Daily Star*, Bangladesh, *The Guardian* from the Maldives, partly the coffee stories) or by focusing on investment in eco-friendly technologies (Norwegian *Dagsavisen* on US solar energy, and Uganda's *Monitor* on biogas). Finally, in the lengthy reportage from Uganda, where local action represents innovation, a win-win situation put in a larger, holistic context, proves it is possible to relate scientific perspectives to peoples' livelihoods and everyday experiences, not to mention that this is possible in a country where print media are scarce and not that resourceful.

Interestingly, the observed spectrum of discourses in these stories of good practices range from grassroots orientation to national interest, from hope and technological optimism to climate justice, from alarming perspectives to scientific uncertainty and, lastly, from citizen and consumer interests to responsibilities. Storytelling about abstract and highly complex scientific topics often requires personalization, visualization and domestication in order to bring the stories into people's living rooms. Climate change coverage in this sense represents no exception. Giving climate change faces and voices can help media efforts to enhance public understanding of the most challenging global problem on their agenda, and citizens in turn may become more aware of the need for action in everyday life.

Furthermore, it would be instructive in future research to expand the range of criteria for good practices in climate reporting, *nota bene* to be understood as quality journalism, in dialogues with experienced journalists from the "climate beat". This should be combined with larger quantitative research on climate coverage in general, mapping voices and genres, texts and visuals. Climate literacy is a challenging issue since in some so-called vulnerable countries there was limited coverage of the IPCC and climate change in general (see Chaps. 4 and 10, Isbrekken 2013). This finding raises the question of whether or not the discourse of vulnerability is found more frequently in journalistic coverage of climate change in the "global North" than in the "global South". Findings drawn from this kind of research could enrich journalism education and the training of climate journalists in particular: experimenting with laboratory and in-house seminars, co-teaching/co-learning and interactions with climate scientists.

The way in which the IPCC—and journalistic media—visualize the IPCC reports, is a separate issue, not treated much here, although visualization is a powerful way of providing audiences with intellectual access to climate change issues. On the other hand, visual stereotypes related to global warming can also have an adverse (alarmist) effect (Hahn et al. 2012).

Good practices in climate journalism depend a great deal on whether or not climate science journalism will itself become a specific job profile in the future. Presently, the journalists covering climate change have many different assignments and backgrounds (Brüggemann and Engesser 2014). Alternatively, climate journalism might become a rather normative concept such as, for instance, the much discussed peace journalism or development journalism (Chap. 10). In this context, further quantitative and qualitative surveys among journalists about their perceptions of professional climate science journalism would be useful.

The IPCC scientists interviewed emphasize the need for specialization and a closer relationship between scientific and journalistic fields, as well as the need for creativity and editorial backing. They have high expectations for a journalistic field currently undergoing profound transformation, where specialized climate reporting is more likely to occur in niche rather than mainstream media. Or, on a more optimistic note, media executives could prioritize the global challenge, if necessary, at the cost of other areas of coverage.

The stories presented in this chapter are all traditional print media products. However, in the last few years, new ways of storytelling within legacy media have emerged. One example is the interactive transmedia storytelling project "Snow Fall" [25] initiated by *The New York Times*. These creative and informative multi-perspectival projects, albeit subject to resource constraints, could improve the quality of climate change journalism in the years to come.

[25] For more on this project visit: http://www.nytimes.com/projects/2012/snow-fall/#/?part=tunnel-creek

BIBLIOGRAPHY

Austgulen, M. H. (2012). *Nordmenns holdninger til klimaendringer, medier og politikk.* Prosjektnotat 4-2012. Oslo: SIFO.

Banda, F. (2008). Deconstructing media coverage of development. *Media Development, 55*(1), 30–34.

Bauer, M. W. (2009). Editorial. *Public Understanding of Science, 18,* 378–382.

Bech-Karlsen, J. (1996). *Ubehaget i journalistikken.* Oslo: Cappelen.

Beck, U. (2010). Climate for change or how to create a green modernity? *Theory Culture & Society, 27*(2–3), 254–266.

Benson, R., & Neveu, E. (2005). *Bourdieu and the journalistic field.* Cambridge, MA: Polity Press.

Boykoff, M.T., & Boykoff J.M. (2004) Balance as bias: global warming and the US prestige press. In *Global Environmental Change* 14, 125–136. Elsevier.

Brüggemann, M., & Engesser, S. (2014). Between consensus and denial: Climate journalists as interpretive community. *Science Communication, 36*(4), 399–427.

Carey, J. (Ed.). (1989). *The faber book of reportage.* London: Faber & Faber.

Carvalho, A. (2005, August 27). *"Governmentality" of climate change and the public sphere.* University of Minho Repository. Retrieved from http://repositorium.sdum.uminho.pt/handle/1822/3070

Eide, E. (2011). *Down there and up here: Orientalism and othering in feature stories.* New Jersey: Hampton Press.

Eide, E., & Kunelius, R. (Eds.). (2012). *Media Meets Climate.* Gothenburg: Nordicom.

Festinger, L. (1957). *A theory of cognitive dissonance.* Stanford, CA: Stanford University Press.

Hahn, O., Eide, E., & Ali, Z. S. (2012). The evidence of things unseen: Visualizing global warming. In E. Eide & R. Kunelius (Eds.), *Media meets climate: The global challenge for journalism* (pp. 217–242). Gothenburg: Nordicom.

Hansen, J. (2014). James Hansen speaks truth to power. *The Des Moines Register,* October 11. Retrieved from http://www.desmoinesregister.com/story/opinion/columnists/iowa-view/2014/10/11/james-hansen-climate-change-speaking-truth-power/17118625/

Hartsock, J. (2009). The "other" literary journalism. *Genre, 42*(1–2), 113–134.

Heinonen, A., & Luostarinen, H. (2008). Reconsidering "journalism" for journalism research. In M. Löffelholz & D. Weaver (Eds.), *Global journalism research. theories, methods, findings, future* (pp. 227–239). Oxford: Blackwell.

Hulme, M. (2014). *Can science fix climate change? A case against climate engineering.* Cambridge: Polity Press.

IPCC. (2012). Field, C. B., Barros, V., Stocker, T. F., Qin, D., Dokken, D. J., Ebi, K. L., Mastrandrea, M. D., Mach, K. J., Plattner, G.-K., Allen, S. K., Tignor, M., & Midgley, P. M. (Eds.), *Managing the risks of extreme events and disasters to advance climate change adaptation. Summary for policymakers.* Cambridge: Cambridge University Press.

IPCC. (2014, March 31). *IPCC report: A changing climate creates pervasive risks but opportunities exist for effective responses.* Geneva, Switzerland: IPCC. Retrieved from http://www.ipcc.ch/pdf/ar5/pr_wg2/140330_pr_wgII_spm_en.pdf

Isbrekken, A. (2013). Klimajournalistikk i stormens øye. Journalistisk dekning av klimaendringer på Salomonøyene. Master thesis, University of Oslo, Oslo. Retrieved from https://www.duo.uio.no/handle/10852/38763

Lakoff, G. (2010). Why it matters how we frame the environment. *Environmental Communication, 4*(1), 70–81.

Myers, A. (2014). "Skeptics" and "believers": The anti-elite rhetoric of climate change skepticism in the media. In L. Lester & B. Hutchins (Eds.), *Environmental conflict and the media* (pp. 261–273). New York: Peter Lang.

Naper, A. A. (2014). Klimakrise eller oljefest? Perspektiver fra kommentariatet. Master thesis, University of Oslo, Oslo.

Painter, J. (2013). *Climate change in the media: Reporting risk and uncertainty.* Oxford, UK: I.B. Tauris and Reuters Institute for the Study of Journalism.

Risbey, J. (2008). The new climate discourse: Alarmist or alarming? *Global Environmental Change, 18,* 26–37.

Said, E. (1994). *Culture and imperialism.* London: Vintage Books.

Secko, D., Amend, E., & Friday, T. (2013). Four models of science journalism: A synthesis and practical assessment. *Journalism Practice, 7*(1), 62–80.

Swyngedouw, E. (2010). Apocalypse forever? Post-political populism and the spectre of climate change. *Theory, Culture & Society, 27*(213), –232.

Vehkoo, J. (2010). *What is quality journalism and how it can be saved.* Oxford, UK: Reuters Institute for the Study of Journalism.

CHAPTER 12

Key Journalists and the IPCC AR5: Toward Reflexive Professionalism?

Risto Kunelius, Hillel Nossek, and Elisabeth Eide

In March 2015, *The Guardian* launched a rare global journalistic campaign entitled *Keep it in the ground*. It aimed at pressuring large investors to draw their investments from fossil fuel and acknowledge that much of the known fossil fuel resources are both morally and financially unburnable. A few weeks into the campaign *Guardian* editor-in-chief Alan Rusbridger wrote that the "rule of newspaper campaigns is that you don't start one unless you know you're going to win it." In a June 2015 research interview, he reflected on the decision to launch the initiative and on the boundary between journalism and advocacy.

> I've been quite limited during my time as editor in actually going out and
> fighting campaigns. Because broadly I do believe in the separation of news
> and comment. But there are some subjects in which it seems to me there's

R. Kunelius (✉)
School of Social Sciences and Humanities, University of Tampere,
Tampere, Finland

H. Nossek
Kinneret College on the Sea of Galilee, Tiberias, Israel

E. Eide
Department of Journalism and Media Studies, Oslo and Akershus University
College of Applied Sciences, Oslo, Norway

257

R. Kunelius et al. (eds.), *Media and Global Climate Knowledge*,
DOI 10.1057/978-1-137-52321-1_12

no reasonable debate. Free speech would be one, female genital mutilation [another]. It's difficult to be pro FGM, and I think on climate change, where the overwhelming majority of scientists agree on something, I think it's forgivable to have a more activist ... I don't know if activist is the right word.[1]

The *Keep it in the ground* campaign and Rusbridger's reflections on it serve well as a point of departure for this chapter, which turns attention to the voices of some of the key journalists who covered the AR5. The example reminds us of *The Guardian's* central role in global environmental journalism: the paper hosts what is likely the world's most resourceful climate reporting team and takes unique approaches to the issue, often appearing as a reference point for other journalists (Eide 2012a). More specifically, Rusbridger's reflections illustrate the obvious need to *justify* the exceptional measures taken in climate advocacy journalism.

Such reflections grow from the assumption that it is imperative for journalism to be somewhat detached from particular solutions to the social problems it covers. Of course, being "objective" or "detached" have always been lofty goals rather than simple realities of journalism, with much local variation in how such professional values have been operationalized in journalistic practice (see, for instance, the discussion on "civic epistemologies" in Chap. 4; Hallin and Mancini 2004). This tension between the ideals that justify journalism and the realities it lives with have produced a rich literature on *journalism as a profession*. In this chapter, we use the debate on professionalism to explore how the specific challenges of climate journalism are affecting the profession. Our case study asks whether, and to what extent, general reporting on climate change and IPCC AR5 reporting in particular articulate a different mode of journalistic professionalism. To answer this question, we interviewed some key climate journalists from around the world, asking them to reflect on their performance in the task of reporting on the AR5.

Professionalism and Its Uses

Inner tensions in the discourse on "professionalism" can be illustrated by identifying two different but interrelated uses of the term. *Materially* speaking, being professional refers to a *set of practices*—a craft. Professionals are identified by what they do and by the skills they master. Historically, modern professionalism can be seen as the continuation of an earlier system of guilds, with masters and apprentices. In this sense, professional journalists

[1] Interviewed in London on June 16, 2015, by Risto Kunelius and Adrienne Russell.

mutually recognize each other by their performance, such as effectiveness in gathering information, skillful use of presentation techniques or mastery of reporting routines under extreme time pressures. A long line of ethnographic research on work in newsrooms has illuminated the ways in which daily journalistic work is organized and reproduced (see, for instance, Tuchman 1978; Boczkowski 2004; Usher 2014). From this perspective, the boundary between journalism and "non-journalism" has always remained porous for two key practical reasons. First, as a set of skills, journalism has been exposed to the volatile forces of modern markets and technology, leading to constant changes in journalistic routines and the evolution of performance demands. The emergence of *blogging*, for instance, highlights the question of what skills actually differentiate a "journalist" from someone who writes actively about current affairs. Thus, pressures for more "participatory," "open," "networked," "hybrid" or "ambient" journalism (e.g. Beckett 2010; Hermida 2010; Chadwick 2013) have intensified. New kinds of collaborative models of journalism are spreading (e.g. Lewis 2012), and social movements around problems such as climate change have been interesting sites for alliances across the traditional boundaries of journalism (e.g. Russell 2013, 2016). Second, differentiated forms of expertise (on economics, sports, culture, sciences, etc.) have for a long time served as a way of entering the field of journalism. Specific journalistic skills often depend on the mastery of expertise or cultural capital outside journalism, leaving journalism as a field somewhat weak from the outset (Bourdieu 2005).

Symbolically, being professional refers to the ways in which professionals *justify* their practices (and by default, their privileges and social position). In the classic "modern" sense, fully developed professions (medicine and law, for instance) have been able to claim they defend the public interest since their work has a scientific basis, demands extensive formal education (sanctioned by the state) and applies an explicitly formulated set of ethical guidelines. In the case of journalism, such demands frequently create disagreements about whether journalism counts as a "proper" profession (Larson 1978; Ahva 2010; Jackson 2010). Despite these theoretical disagreements, journalists themselves (in many parts of the world) identify often with an abstract set of values. Deuze (2005: p. 447), for instance, lists five identity characteristics that journalists often subscribe to: public service, objectivity (neutrality, fairness), autonomy, immediacy and ethics. Such features of the public legitimation discourse are a crucial aspect journalistic autonomy and professionalism. Arguably, journalism as a profession is particularly dependent on this symbolic dimension, as questions about specific skills, about accreditation and education, are more vague and fleeting. As the technological and institutional framework of mainstream journalism has been rapidly changing the future of professional journalism questioned, as eloquently put by Waisbord (2013):

> The apogee of professional journalism is behind us. It sounds anachronistic in a world of amateur journalism and partisan media, when the boundaries of press systems are constantly expanding. Yet as journalism navigates a changing world and confronts doubts about its future, it is important to understand how journalism maintains its unique position in society, how it reinforces its role as the expert arbiter of news as it tries to maintain autonomy vis-à-vis external forces. (p. 10)

A strong public dependency on particular but ambiguous values, such as "fairness," "objectivity" or "transparency," has not always been only empowering for journalists. In climate coverage, Boykoff and Boykoff (2004) illustrated how adhering to the notion of "balance" led to a particular bias in favor of skeptics. Similarly, journalism's commitment to the "transparency" of other social institutions (such as science) no doubt provided fertile ground for "Climategate" in 2009, casting exaggerated doubt on climate scientists (e.g. Pearce 2010).

Neither the material nor the symbolic aspect alone captures what is at stake in professionalism. The core of public accountability professionalism lies in the tension between practices and the discourses that justify them. Climate change challenges us (as Rusbridger's reflection demonstrates) to think about how *particular topics* can shape professionalism: both as a practice and as a justification discourse. This underlines a specific way of thinking about what it means to be accountable to *a public*. "Publics" can refer to many things, ranging from elites to the whole population of nation. Following a pragmatic path, "publics" consist of those who are affected by the action of others and thereby have a need to discuss and effectively manage these public problems (e.g. Dewey 1927). As recent theorizations have underlined, global problems and consequences of local action mean that a proper framing of such publics becomes transnational (e.g. Splichal 2009; Fraser 2014; Honneth 2014; see also Chap. 6). This transnational issue-specific notion of the "public" and the consequent "accountability" for it raises the question that frames our analysis of the professional reflections of climate journalists: Can climate change be viewed as a global *problem discourse* that articulates new kinds of *transnational publics?* Could such a global problem discourse serve as a platform on which *transnational professional journalism* could emerge?

GLOBAL CLIMATE JOURNALISM?

The idea that professional journalism is not only the product of the institutional contexts in which it develops (see Hallin and Mancini 2004; Brüggemann et al. 2015) but also effected by the particular, epoch-defining challenges it faces is not a new one. Zelizer's (1993, 2010) notion of journalism as an "interpretative community" carries a similar perspective: that through professional debate and discussion, concrete conjunctures, events and issues are incorporated into the professional fabric of journalism. Hanitzch (2007) has identified several reform movements that have challenged narrow professional culture by highlighting causes worthy of journalism to commit itself to, such as conflict resolution (peace journalism, e.g. Lynch and McGoldrick 2005) participatory democracy (public journalism, e.g. Rosen 1999), or development journalism (Chap. 10). Different justifications for journalists to allow diversion from formal professional criteria have been observed in cases of war, terrorism and other issues of national urgency (Nossek 2004). Recently, Nerone (2015: pp. 222–230) condensed such remarks well by referring to particular problems and historical situations as the "test of capacity" for journalism. Climate change is one of his prime, contemporary examples.

Some studies have indeed suggested that climate change is an issue that provokes changes in the professional culture. Studying seasoned environmental journalists in the USA, Hiles and Hinnant (2014) observed that some journalists "…have developed a process of verification that allows them to use authority or 'opinion' when evaluating sources (though opinion stays out of the final story)" (p. 445). They also found differences among climate reporters, with some more willing to detach themselves from old-fashioned objectivity. Interviewing Swedish climate journalists, Berglez (2011) identified three kinds of professional creativity. Some emphasized the importance of simplifying the complicated messages of science, as journalists should. Others, particularly specialized environmental journalists, tried to incorporate the virtues of science (abstract, complex language) into their journalism. Yet, another group saw climate reporting as a more game-changing issue with the potential to challenge the existing limits of professional journalism. From this perspective, journalistic creativity became the ability to "…let the climate outlook influence the news production as far as editors, chief editors, and media owners allow you to go" (ibid.: p. 461).

Following Beck's (1992, 2009) notion of *risks* becomes a key feature of any given new global climate professionalism. Cottle (2009), for instance, sees that global media construct risks in a manner that can support possible transnational public agendas. Cox and Pezzullo (2015: pp. 159–165) distinguish between technical and cultural ways of constructing risk assessments, suggesting that climate journalism, even when based on the risk frame, can remain within the boundaries of linear science journalism (Secko et al. 2013; Chap. 2). A more cultural notion of risk recognizes the social context and meaning of the risks involved. This could carve out space for broader collaborative relationships among citizens, experts and agencies. Thus, by showing the importance of thinking "knowledge" as "risks", climate journalism could become an integral part of a deeper rethinking of professional journalism's relationship to the production of knowledge about complex problems.[2]

In addition to challenging a core value of journalism (knowledge), climate change transcends the national home bases of knowledge production. In Beck's horizon (2010), such a "cosmopolitan," journalism would demand a *global* transnational strategy that transcends narrow nationalism. Developments toward such "global journalism" have been sketched out in several ways. Reese (2008) describes it as a transnational system for collecting, editing and distributing material that creates a "global news arena" where "multiple perspectives and interpretations are in circulation at any given time, but still with a heightened mutual awareness, reflexiveness, and timeliness in their reaction to one another" (Reese 2008: p. 242). Cottle (2011) underscores the potential of transnational crises or events (see also Kunelius et al. 2007; Eide et al. 2008; Nossek 2004). Berglez (2007, 2008, 2013), in turn, defines global journalism through transnational interdependence (capital, trade and human mobility) as coverage that "investigate(s) how people and their actions, practices, problems, conditions, etc. in different parts of the world are interrelated" (2007: p. 151). Global journalism, from this perspective, would be news that portrays complex relations which cut across organizations (UN, NATO), identities (transnational businesses/middle class/working class) and institutions (political/economic); journalism that concentrates its reporting on problems, power struggles and politics at the *trans*national level.

Recent empirical research on journalists who report on climate change offer further insights into the professional developments we focus on here. Brüggemann and Engesser (2014) surveyed journalists in five countries (Germany, India, Switzerland, the UK and the USA). Despite the differ-

---

[2] For a lengthier reflection on risk and climate journalism see Painter (2013, 2015b).

ences between their local contexts, common patterns appeared in interpretations of climate change: the assessment and handling of climate skeptics, expertise in climate coverage, and the predominant use of sources from the scientific community. Their findings suggest that climate change journalists can be viewed as members of an interpretive community that overlaps with the scientific community, pointing out that climate science could serve as a resource for global professionalism. Elaborating on the cognitive frames of journalists, Engesser and Brüggemann (2015) point to other interesting differences. There are signs of a global "blame game" between journalists from developed and developing countries. They also detect an unwillingness among journalists from the Global North to view climate change as a radical challenge to capitalism. Instead, they emphasize the need for a better global "ecological discourse" where communication is a key factor in finding solutions.

The AR5 opened an entry point to examine the question of whether the practice of covering climate change in general and the AR5 in particular follows common journalistic professional norms or whether there are signs of an emerging transnational professional practice. In order to do so, we interviewed a select group of key climate journalists, each active in shaping the AR5 coverage in different countries. An open, in-depth questionnaire was designed to probe answers to the research questions and administered to the journalists via email, skype and in person. The questionnaire was answered by 18 leading climate journalists from 12 countries (Sidebar 12.1).

The professional experiences of the respondents reflect a reasonable diversity of locations and media outlets, but by no means provide an exhaustive inventory of all angles. However, together they carry an extraordinary amount of professional experience; their average age was 50

Sidebar 12.1

Interviews with key journalists

The international journalists interviewed for this chapter were selected by identifying those who were most frequently bylined in our sample of IPCC AR5 coverage. In April 2015, 35 journalists were contacted by email. When some journalists were not reached, new contacts were added to the list in order to establish a reasonable global sample. The final sample includes journalists that responded to our interview request by the end of August 2015.

The interviews were conducted using a semi-structured thematic format that: (1) collected basic background information on the career, education and position of each journalist; (2) discussed the usefulness and quality of the IPCC communication from a professional standpoint and the use of other information sources and stakeholders (science, politics, business, civil society, audience); (3) mapped international professional reference points (colleagues and news outlets) and cooperative networks; and (4) reflected on the particular challenges of covering climate change and climate science.

Some interviews were conducted face-to-face (ff), some on the Internet through a live connection (iff), and some in written form through email-contact (email).

List of respondents (interviewer and interview method)

Marie-Noëlle Bertrand, *l'Humanité*, France, 8 July 2015 (RK, ff)

Stéphane Fourcart, *Le Monde*, France, 9 July 2015 (RK, ff)

Tony Carnie, *The Mercury*, South-Africa, 5 May 2015 (RK, email)

Damian Carrington, *The Guardian*, UK, 10 July 2015 (RK & EE, ff)

Angela Davidova, freelance, *Kommersant*, Russia, 30 April 2015 (RK, iff)

Justin Gillis, *New York Times*, USA, 23 June 2015 (HN, ff)

Mark Hannam, *Sydney Morning Herald*, Australia, 17 May 2015 (RK, email)

Toru Ishii, *Asahi Shimbun*, Japan, 9 May 2015 (SA, email)

Ole Mathismoen, *Aftenposten*, Norway, 28 April 2015 (EE, skype)

Matti Mielonen, *Helsinkin Sanomat*, Finland, 27 April 2015 (RK, email)

Ifterkhar Mahmud, *Daily Prothom Alo*, Bangladesh, 21 August 2015 (RK, email)

Brigitta Isworo, Laksmi, *Kompas*, Indonesia, 4 April 2015 (RK, email)

Andrew Revkin, freelance, *The New York Times*, US, 7 July 2015 (RK & EE, ff)

Zafrir Rinat, *Haaretz*, Israel, 1 April 2015 (RK & HN, email)

Tor Sandberg, *Dagsavisen*, Norway, 26 April 2015 (EE, email)

Daisuke Sudo, *Asahi Shimbun*, Japan, 9 May 2015 (SA, email)

EE = Elisabeth Eide, RK = Risto Kunelius, HN = Hillel Nossek, SA = Shinichiro Asayama years, and all of them have been journalists for over a decade, some for well over 30 years. Most have been following the topic for several years or decades. They are well established in their newsrooms, and each told that they enjoyed considerable professional freedom in deciding what to cover and how. There are few freelancers in the sample, but they regularly work for key media. Typically there are two routes to a key climate reporting position that the interviewees represented. Many had an educational background in the natural sciences (geology, physics, etc.) but had chosen to venture into journalism instead of an academic career; often supporting this with a journalism degree later. Others gained their cultural capital through a more generic career in journalism, serving in foreign news sections, political reporting or other fields.

In what follows we mostly try to *listen* to the voices of this professional community, hearing what kind of concerns they raised about the AR5, the IPCC and climate reporting more generally. The analysis is divided into four sections: (1) views on the IPCC and its communication practices; (2) reflections on the craft, rules and routines of the journalistic profession; (3) the dynamics and boundaries of the interpretative community (relationships among professional journalists and between them and other stakeholders); and (4) specific challenges, and the possible uniqueness of climate change as a global problem.

COMMUNICATING THE IPCC AR5: A MIXED JUDGMENT

Overall, the journalists in our sample had a relatively favorable view of the IPCC. Judged against its hugely complex task, they often saw merit in the work of the panel with one journalist calling it "…the most peer reviewed document in the history of humanity." What was appreciated throughout, sometimes explicitly, was the long trajectory and work of developing (a quantified) language that has shown the increasing certainty of the evidence. Several journalists also pointed to enhanced IPCC communication efforts between AR4 and AR5. The *Summaries for Policy Makers* (SPM) were generally deemed useful, although many said that they were not ideal for helping to communicate the results to a wider audience.

> I found them useful, but I'm already someone specialized in the topic and I'm already someone who knows a lot. I don't know, maybe you should also speak to someone who has less experience in the issue. (Angela Davidova, freelance, *Kommersant*, Russia)

While journalists recognized that the SPMs are not meant for the general public they still maintained that further effort could have helped make the "information more accessible." In general, more clarity about the main points, more concrete language and more (and better) graphic elements were in demand.

> ...the IPCC has learned during the years to use more simple and concrete language. However, the report language could still be more concise and short. For example, formulating a short summary with the relevant data. Current executive summaries are still overloaded with too many terms, definitions, and statistics. (Zafrir Rinat, *Haaretz*, Israel)

Journalists from Bangladesh, Japan and Indonesia added an obvious but important point: that key global materials, like the SPMs, should be translated into at least some of the major languages. (Bengali, for instance, is spoken by some 250 million people in Bangladesh and India)

While many journalists tended to soften their critique by reflecting on the difficulty of the IPCC communication task, there were also some sharper moments of criticism.

> I would describe it as a disaster as an exercise in public communication (...) it's written by a committee and reads like it's written by a committee. (...) they do try but the rules [of the IPCC] essentially prevent the PR staff from doing this in a way that would achieve effective coverage. The mere fact that the thing is not finalized until a big meeting and then it's released almost immediately, prevents the normal embargo system that you have in most types of science coverage which allows people [journalists] to do a better job. (Justin Gillis, *New York Times*, US)

In addition to demands for a more effective service for reporters, criticism toward the IPCC also extended to the level of scientific evidence. Several journalists saw the AR5 message as too conservative and cautious. The IPCC was "overestimating their own uncertainty" by downplaying "the upper range of possibilities" or "artificially" trying to "emphasize what is positive:"

> They write about the 2 degree aim as if it is possible without requiring a revolution. They talk about a green shift, but I think what is happening is more a green adjustment. (...) There is a divergence between the report itself and the SPMs. But they cannot deviate in writing from the main con-

clusions, although the dramatic sides are somewhat hidden. They are a bit UN-ish-positive. (Ole Mathismoen, *Aftenposten*, Norway)

Reporters linked the cautious tone of the IPCC to the fundamental role of uncertainty in science ("scientists will be sure of a disease only after the patient is dead") but also to the political imperative of the IPCC consensus process ("remember that Saudi Arabia is in the room"). On the other hand, some opposing voices suggested that the IPCC had become *too radical* and more *alarmist*, highlighting "the worst case, instead of the full sweep of possibilities." We will return to this difference of views below.

THE LOGIC OF JOURNALISM: A REFLEXIVE SELF-CRITICISM

While the practical realities of news work justified specific demands for changes in the IPCC communications, seasoned professionals also pointed out the problems embedded in their own routines. This reflexivity was, at times, somewhat defensive, emphasizing the circumstances and constraints in their reporting.

...the amount of information in the reports makes it quite difficult to fully digest the reports in a day and before the deadline. I usually have to ask scientists to comment on the news today and later on I may try to check out some facts from the report itself (Matti Mielonen, *Helsingin Sanomat*, Finland).

Ottmar Edenhofer [chair of WGIII] was good, "It doesn't cost the world to save the planet." That's kind of handy. (...) If you think about what happens on those days, you've got a thousand-page report, you've got a 50-page Summary for Policymakers. You've got, I don't know, four hours to file, maybe? You're not going to phone round 50 people on the off-chance they might have something interesting to say. (Damian Carrington, *The Guardian*, UK)

These moments show professionalism as a set of necessary routines that can be applied to any topic, a set of practices justified by the pressures of time and space. On the other hand, journalists were conscious and critical of mere "stenographic" reporting and attempt to reduce climate change to "sexy, oversimplified sound-bytes."

Another common concern was the competition for space inside the daily news hole, the "tyranny of front-page thought":

> ...[by] middle of the afternoon, the science editor has to go into the front-page meeting and say, "This is really important because..." but you're competing with stock market scandals, Greece, et cetera, et cetera, ISIS. (Andrew Revkin, freelance, *New York Times*, US)

Reflecting on the struggle to avoid getting "relegated to the inside pages or not used at all," several climate journalists pointed to the benefits of being situated in a specific science or environmental section. Such "dedicated space" helped them to carve out more local, professional autonomy. Moreover, the web-versions of special sections offered a way of getting stories "out there," even if the battle for attention in the seemingly more prestigious print version was lost.

> We run our own team. Online, we can publish whatever we want, basically. To get into the newspaper, of course, that requires other conversations, because there's an editor of the newspaper, and we have to fight for space there. (Damian Carrington, *The Guardian*, UK)

However, these organizational solutions were viewed as a mixed blessing, separating climate into a "niche" category instead of "penetrating all journalism."

> ...political journalists don't care enough about climate change and the issues of climate change. (...) They still think that it's a secondary issue. Things [will] begin to change because civil society and the politic themselves are beginning to speak more about climate and about the transition. (...) They [will] begin to think that there is something quite important. (Marie-Noëlle Bertrand, *L'Humanite*, France)

Another often repeated and partly self-critical point from several journalists was that the IPCC and climate science have already delivered their important message several times. This has placed journalists and policy makers in a situation where "...there is an increasing sense of fatigue around climate change news that simply restates previous challenges and dilemmas (among both readers and journalists)."

Beyond the issues of space and repetition (what is new about the climate story?), journalists struggled with the non-event nature of this topic; a characteristic that effected newsworthiness and moved the IPCC story to the inner pages. Finally, the uncertainty associated with different risk assessments had implications on the number and salience of climate change stories in many newsrooms.

ROLES AND RELATIONSHIPS: BOUNDARIES OF THE INTERPRETATIVE COMMUNITY

When talking about other journalists, scientists, NGO-activists and other stakeholders in climate communication, the interviewees provided insights into the boundaries of their particular interpretative community. Their reflections on these interfaces point to some specific characteristics of professionalism articulated around global climate (science) journalism.

Inside the field of journalism (between journalists), respondents frequently refer to the traditional professional virtues of independent reporting and mutual competition. However, there are also moments of heightened transnationalization. This is evident among the leading institutional and individual voices in climate journalism. *The Guardian*, *The New York Times* and occasionally *Reuters* are frequently mentioned as self-evident, leading reference media. This draws attention to a joint agenda building network where some outlets, with their superior investments of resources, are clearly benchmarks for reporting. This suggests a new, intensified role for such media outlets in the global flow of news content and story angles.

A transnational community of professional actors was also visible in the interviews, demonstrated by a shared international recognition of certain individuals. Clearly, some journalists have made a reputation for themselves that carries weight among their international peers. This fame advantage favored those who published their work in English. While something may be qualitatively new about the transnational aspects of this evidence, it still corresponds with standard arguments about a sense of professional solidarity. This sense of respect did not translate into practical relationships *between* these professionals. Individual cooperation across borders was an alien idea to most respondents, who viewed peer relationships competitively. There were, however, some examples provided of institutional and organizational level cooperative networks.

> On the ground at a conference [for example], we just do our own thing, if we're there. But we actually have a lot of collaborations already. We have *The Guardian Environment Network*, which has about 20 publications, and

we content-share with them. They run our stories, we run their stories like *BusinessGreen* and *Yale 360* and a bunch of others. (...) We're really happy to share. Partly because we're owned by charitable trust, so we don't have the same commercial imperative. Although, we run on a commercial basis. (...) we have a network of bloggers now who work for us around the world. They're not on staff, but we run their copy. Actually, just as part of that came out of the *Keep it in the ground* campaign, there's a *Climate Publishers Network*, I think we've called it now which involves *Le Monde* and a bunch of other newspapers. Collaboration is absolutely key. We learnt that in other ways from *WikiLeaks*, for example. (Damian Carrington, *The Guardian*, UK)

Hence, despite some traditional competitive ethos, we can also see the emerging shape of a network of collaborative platforms. Mainstream agenda-setting outlets (such as *The Guardian*) are linked to more niche publications and individual bloggers. Journalists, likewise, are working within a partially shared global infrastructure of intermediary platforms of climate knowledge (such as *Carbon Brief, RealClimate, Climate Progress* or *RTCC*).

Reflections on the boundary between journalists and global *climate science* produced mixed reactions. As the talk of "stenographic" journalism reveals, a traditional division of labor between reporters and their sources is still viable. In this respect, the availability of local and domestic scientific sources proved to be a crucial resource for journalism (see also Chaps. 2 and 3). Highlighting this dependency, an Indonesian journalist described local sourcing challenges, explaining they are "difficult to find as there was only one researcher from Indonesia Meteorology Climatology and Geophysics Agency who followed and got involved in the AR5 writing process" (Brigitta Isworo Laksmi, *Kompas*, Indonesia). This poses a major problem for journalists from countries where IPCC authors are scarce or non-existent. As an Australian journalist iterates, the consequence is that: "too little is written about the climate changing in poorer nations. They tend not to have the scientific communities the close meteorological watch that rich countries have" (Peter Hannam, *Sydney Morning Herald*, Australia).

Other respondents felt competent enough to go beyond the traditional routines of reporting through sources. Building on their science background or long experience in journalism, they were ready to become more engaged in the substantial issues at hand. As one journalist put it: "Well, natural science is rather difficult, but not *that* difficult." Such views often meant taking distance from other "mainstream" journalists and identifying more with the scientists.

Even inside the scientific community you have people (...) that have thoughts about climate change which are completely wrong. Mainstream journalists tend to think that someone with a science job, professor in the university, a researcher (...) [is] always saying something more or less right about anything that deals with science in general. As a science journalist, when you dedicate all your time to study the problem from a scientific point of view, you're able to counter that. (Stéphane Foucart, *Le Monde*, France)

In addition to becoming an active *interpreter* of science this specialized expertise could stretch to more practical, interactive ways of verifying evidence.

...if a paper was coming out or the IPCC had a section that I thought was dubious, I would send it to 5 or 10 people who are like my own peer review. (...) I know enough sea ice scientists, enough glaciologists, enough computer modelers, enough tropical forest researchers that if a paper is coming out I'll send an email to 5 or 10 people and I'll say "Please let me know what you think of this." (Andrew Revkin, freelance, *New York Times*, US)

Such moments suggest that some journalists are situated not only within the field of journalism but also at the fringes of the field of climate science. Hence, while some journalists identified and were sometimes frustrated by the gap between science and journalism, they also demonstrated an understanding of the limited role of scientists.

There is a huge problem here, sort of a fault line, between journalism and science in the sense that when an event is hot, when public attention is focused on particular disasters, we want answers from the scientists right then about, "OK, what's the role of climate change in this event?" They can't give us those answers. It takes them two or three years to analyze. By that point, the public attention has moved on. (Justin Gillis, *New York Times*, US)

A particularly interesting interface for developing the interplay between these two fields was the demand for more rapid and effective, but scientifically valid interpretations, of the linkage between extreme weather events and climate change.

I am expecting that from now on if the method of 'event attribution' can demonstrate continuously the relevance of climate change and extreme weather events, it may possibly keep our eyes open and maintain the sense of crisis. (Daisuke Sudo, *Asahii Shimbun*, Japan)

Respondents also discuss the *role of NGOs* and civic activist groups in helping to interpret IPCC knowledge.

> ...quite a few local NGOs who worked on environmental issues (...) take up the facts and build up their own press release based on the IPCC report. (...) WWF does the same. They add up more information on Russia (...) the interpretation comes from two sources. One is NGOs and another one is (...) the Russian State Hydrometeorological service. They also publish similar assessment reports for Russia regularly. (Angela Davidova, freelance, *Kommersant*, Russia)
>
> Civil society activists are the most fluent in discussing climate change issues, in adaptation as well as mitigation issues. Many of them have an academic background and they are backed with academic data and research. Yet, in those climate change issues, what is talked about and discussed by civic activists is rarely heard as an input for policymakers even though they are invited to get involved in the process. (Brigitta Isworo Laksmi, *Kompas*, Indonesia)

The role of NGOs is particularly interesting in the IPCC case, as the quote above illustrates. NGOs are situated between the IPCC (who by the nature of its mission is cautious) and journalism (who by nature of its mission hunts for effective narratives dramatizes new information, etc.). This role will remain important even if the IPCC's enhanced efforts to communicate more effectively are bringing some solutions to the practical gaps of communicating climate knowledge (Chap. 3). From the standpoint of professionalism, there is some interesting evidence of "NGO-journalism" that tests and broadens the professional limits of journalism (Powers 2016).

Reflections on the *audience interface* once again point to traditional roles and new approaches. Direct audience feedback was not discussed much in the interviews and did not seem a crucial factor effecting interviewees. Feedback was occasionally described as "minimal," and sometimes swept aside as badly informed or irrational.

> I get the occasional hostile email pointing to heatwaves in Australia and it's not warmed in 18 years, etc. feedback. Fairfax Media has a strict policy of editorial independence, so if there are stakeholders unhappy with my coverage I wouldn't be influenced by it. Long may that be the case...! (Mark Hannam, *Sydney Morning Herald*, Australia)

More interestingly for the shifting profession, some journalists also expressed new ways of interacting with their audience. For several of our respondents, the traditional media outlets were only one way reaching out, and they described active strategies of utilizing the networked media environment to build and to interact with the readers. *The Guardian's* strategy represents an institutionalized version of this interactive and open approach, but individual journalists took these initiatives too.

> I also actually do quite a lot of PR for my articles. So I post it everywhere, on Facebook, I post it in specialized groups, I send it in to environmental mailing lists, and then I get quite a lot of positive response, I would say. (…) I'm a very active Facebook user. I have something like 5,000 friends. So I act, myself, as a mini media, basically in Facebook. I also post (…) to most environmental and climate groups. I send it to environmental mailing lists which are huge … in Russia there are huge environmental mailing lists, with a few thousand people who are subscribed. So I'm trying to raise up the relationship. (Angela Davidova, freelance, *Kommersant*, Russia)
>
> …an issue like climate is way too big for anyone to cover as a traditional beat. (…) I respond a lot to what I read elsewhere, and critiques. On a complicated issue the best way for an audience to gauge reality is to see at least a couple of different voices and to start to understand, "Well, what we know is somewhere in that space between all these people." That's really hard, because most people go into the journalism world to find shortcuts. (Andrew Revkin, freelance, *New York Times*, US)

Thus, in relation to several stakeholders in the field of climate change and politics, journalists recognize possibilities and new practices. This does not mean that they would give up their *distinct* role in much of this inter-action and collaboration. However, given the weight of climate change as an issue, the boundary reflects important signs of how a particular problem discourse, such as climate change, can bring new reflectivity to the profession.

PROFESSIONAL EXCEPTIONALISM AND THE CLIMATE BEAT

Finally, respondents were explicitly challenged to think about whether reporting on climate change was in any way a different or specific field of journalism; whether there were aspects of fact verification, ethical consid-erations or other factors that demanded exceptional professional concerns. Here, reflections often took an interesting rhetorical form. First, respon-dents would broadly deny any specific nature of the topic, but they would later go on to address aspects of climate journalism that at least make it somewhat different. Thus, by initially declaring a general loyalty to jour-nalism, their core professional community and identity, the respondents liberated themselves. This allowed them to offer interesting insights into the specific pressures they face as climate journalists. The list of partly overlapping pressures and challenges of climate reporting is diverse.

Several journalists said the demands of climate reporting included being "extra careful," paying attention "not to make any value judgments" because "assertive description can give the excuse to climate deniers to attack." Pressure from different lobby groups also had an effect on the media landscape, as in the case of Australia.

> Climate change issues are probably not that special. They require the same reporting skills that would apply to most other issues. The difference, perhaps, is that in some countries (e.g. Australia) there is a well-funded counter-lobby that has willing media outlets (one in particular: Rupert Murdoch's *News Corporation*) to sow seeds of doubt and undermine the public's willingness to respond. (Mark Hannam, *Sydney Morning Herald*, Australia)

Many journalists referred to the problems of global justice and inequity associated with climate politics, stressing that these ethical aspects of the coverage made the work more complex. Our both Japanese respondents pointed out that geographical proximity underscored their solidarity toward the peril of the "small island states." Similarly, local and everyday-experience triggered ethical concerns in journalistic reflections.

> In Indonesia we have many examples of how people live with climate change threats. There are communities in the forest areas, coastal areas, also in the river banks who intelligently applied the principles of living with nature in sustainable ways. There are examples of zero emission farming in local communities demonstrating that people can use forest resources as wisely as indigenous people do. Unfortunately, the policymakers seems never learn from these practices by mainstreaming it into a national policy for adaptation and mitigation. The ethical consideration becomes important when I write about a poor local community or indigenous peoples who defend their resources in a sustainable way. These groups have never been asked to participate in the policy talks, writing about their situation from a mainstream perspective is not proper. (Brigitta Isworo Laksmi, *Kompas*, Indonesia)

Another aspect of this complexity was the unusual timescale of climate politics (see also Chap. 1): the "issues are likely to play out over the decades" and this "means current taxpayers and consumers can see little urgency to respond;" moreover, "you do not win an election by solving something which will happen (much) later."

Few journalists explicitly linked *professional* challenges to the global or transnational nature of the problem, but it was not difficult to see that their detailed awareness of the scale of the problem framed how they understood their own roles as journalists. We let three different kinds of moments in the interviews speak for themselves.

> I do believe that journalists should be in the front of equality and justice issues to get a better world that can answer the climate change issues (…), we don't belong to any government or policymaker. We belong to humanity. I do hope journalists from all countries can share this responsibility. (Brigitta Isworo Laksmi, *Kompas*, Indonesia)
> Those who have really cared to inform themselves in this field know that humanity is facing a major crisis, and the time span might not be too long, also when it comes to biological diversity, health, sea level, etc. (…) Of course I have a journalistic agenda; if not I would not have been working for 30 years with this as my main guiding issue. Climate change is very important, very frightening. I may contribute a tiny little bit to improvement. But as a worker in the profession, I abide by the professional rules. (Ole Mathismoen, *Aftenposten*, Norway)
> Just considering the stakes. You have a really complicated science. The fate of the Earth is in the balance. The amount of journalism expertise that we have on the subject is pitiful, in comparison, right? (Justin Gillis, *New York Times*, US)

Finishing our analysis with such relatively bold declarations of the exceptional nature of the climate challenge and its influence on journalism, it is important to balance the reading. At the same time as journalists recognize these challenges, many of them also articulate more classical virtues and values of the professional field (such a critique of scientist's language and slowness or reliance on the journalistic competition). What makes the intersection of climate science and politics interesting in this sense is that these both views can coexist. That in itself can be taken as an indication that public discourse about climate has already had some effect on professionalism.

CONCLUSION: RESOURCES OF REFLEXIVITY

Our broad concern in this chapter has been to examine the IPCC AR5 communication process from the perspective of the professional journalists involved in it. Mindful of the limited sample of journalists interviewed on one hand, but their leading role in the coverage on the other, we want to tease out some reflections on the emerging issues of professionalism that were introduced at the beginning. Three overlapping themes are relevant.

First, the interviews underscore the need to understand and appreciate professionalism as demanding reflexivity about the rules, practices and values of a given field. The interviews demonstrate that journalists are well in command of their traditional routines and practices. They see and appreciate the values embedded in these skills. When describing their reporting practice and their professional norms most of the respondents articulated a basic, Western conception of journalistic professionalism (accuracy, detachment and precision). In one sense, then, professional journalism seems well intact and, in the eyes of the journalists, quite well equipped to tackle climate science and politics. At the same time, however, we see how experienced climate journalists critically reflect on their own position and—crucially for our argument here—how these critical reflections are linked to the specific qualities of climate coverage.

Second, the interviews suggest that a key source for critically rethinking journalism, among the interviewees, was the interplay of the fields of climate science and journalism. The professionalism of our respondents is shaped by the general field of journalism and the intense interaction with other stakeholders in climate politics. On the one hand, we see journalists naturally skillful in articulating the specific rules of their own profession (demanding more clarity from the IPCC, criticizing its communication practices, struggling inside newsrooms for space, competing with colleagues and so on). On the other hand, when interacting with climate change stakeholders, climate journalists must grapple with the complexity of scientific knowledge, while resisting myopic political interests and simplistic public interpretations. They must be mindful of the complexity of science and the complicated nature of climate politics—but they are also able to highlight the problematic consequences of these things and criticize the IPCC and climate politics for their obvious shortcomings. Similarly, these journalists are critical of their own professional routines and some are eager to innovate with new modes of journalism. In this critique, their knowledge and networks of climate science and politics function as a resource for broadening the profession. Thus, being situated in the borderline between journalism and an overarching problem discourse that links them to other stakeholders is a source of inspiring reflexivity.

Our third point is to focus on the boundary between the national versus the transnational. Here, a dose of conservatism is crucial at first: we have acknowledged throughout the book that the structural forces of national domestication powerfully affect local articulations of the journalistic field. However, our interviews with key journalists identify some signs of the potential to cross over this traditionally strong professional boundary. To begin with, it is clear that climate journalists are partially situated in transnationally shared networks of information. General science forms part of this transnational network, with many journalists citing key scientific journals as important sources of information. The IPCC is another obvious example, but so are international NGOs and other intermediary actors that allow journalists to cross boundaries, providing elements of a shared news agenda concerning what is important. In terms of professional roles, most of the respondents can be characterized as educators—whether their target is the public at large or legislators and governments. Implicitly, they speak in the name of a broader community of interest that extends from journalism to sources of knowledge (the IPCC, climate science and NGO actors). Many identify collaborations with NGOs for a global cause in the service of humanity: "saving the world from itself." Sometimes this makes for activism, but often only in a form that is professionally justified. More specifically, our interviews point to emerging signs of a transnational field of recognition between peers in journalism. These reference points function at the level of individual journalists and at the institutional level (newsrooms).

The Guardian's Keep it in the ground campaign provides an example of climate journalism taking an active role in "saving the world from itself." For some journalists, this campaigning surely represents an unacceptable transgression of professional boundaries. For others, it represents an attempt to speak in the name of a "public" larger than the one with which governments and legislators are familiar. Both of these perspectives are articulations of professional autonomy—journalists trying to retain control over their own work. In the former version, journalistic professionalism seeks its autonomy by protecting its boundaries and bracketing some practices off as illegitimate. In the latter version, journalistic professionalism expresses its autonomy by choosing to advocate for an issue, justifying this action by the exceptionality of the case at hand. This tension between traditional, modern professionalism and the new problems journalists face in an interdependent, globalized world was richly present in the reflections of key journalists who covered the AR5. The nuanced reflexivity exemplified by journalists in the chapter suggests exciting and diversified roles in which journalists function as part of the global flows of climate knowledge. It also suggests an inspiring future for journalism in a world plagued by complex global problems.

BIBLIOGRAPHY

Ahva, L. (2010). *Making news with citizens: Public journalism and professional reflexivity in Finnish newspapers*. Tampere: Tampere University Press.
Beck, U. (1992). *Risk society: Towards a new modernity*. London: Sage.
Beck, U. (2009). *World at risk*. Cambridge, UK: Polity.

Beck, U. (2010). Climate for change or how to create a green modernity? *Theory Culture & Society, 27*(2–3), 254–266.

Beckett, C. (2010). *The value of networked journalism.* London: London School of Economics and Political Science.

Berglez, P. (2007). For a transnational mode of journalistic writing. In B. Høijer (Ed.), *Ideological horizons in media and citizen discourses: Theoretical and methodological approaches* (pp. 147–161). Göteborg: Nordicom.

Berglez, P. (2008). What is global journalism? *Journalism Studies, 9*(6), 845–858.

Berglez, P. (2011). Inside, outside, and beyond media logic: Journalistic creativity in climate reporting. *Media, Culture & Society, 33*(3), 449–465.

Berglez, P. (2013). *Global journalism.* New York: Peter Lang.

Boczkowski, P. J. (2004). *Digitizing the news: Innovation in online newspapers.* Cambridge, MA: MIT Press.

Bourdieu, P. (2005). The political field, the social science field, and the journalistic field. In R. Benson & E. Neveu (Eds.), *Bourdieu and the journalistic field* (pp. 29–48). Cambridge: Polity Press.

Boykoff, M. T., & Boykoff, J. M. (2004). Balance as bias: Global warming and the US prestige press. *Global Environmental Change, 14*(2), 125–136.

Brüggemann, M., & Engesser, S. (2014). Between consensus and denial: Climate journalists as interpretive community. *Science Communication, 36*(4), 399–427.

Brüggemann, M., Engesser, S., Büchel, F., Humprecht, E., & Castro, L. (2015). Hallin and Mancini revisited: Four empirical types of western media systems. *Journal of Communication, 64*(6), 1037–1065.

Chadwick, A. (2013). *The hybrid media system: Politics and power.* Oxford: Oxford University Press.

Cottle, S. (2009). *Global crisis reporting journalism in the global age.* Maidenhead, England: Open University Press; New York: McGraw-Hill.

Cottle, S. (2011). Taking global crises in the news seriously: Notes from the dark side of globalization. *Global Media and Communication, 7*(2), 77–95.

Cox, R., & Pezzullo, P. C. (2015). *Environmental communication and the public sphere* (4th ed.). Los Angeles: Sage.

Deuze, M. (2005). What is journalism? Professional identity and ideology of journalists reconsidered. *Journalism, 6*(4), 442–464.

Dewey, J. (1954). *The public and its problems.* Athens: Swallow Press & Ohio University Press.

Eide, E., Kunelius, R., & Phillips, A. (Eds.). (2008). *Transnational media events: The Mohammed cartoons and the imagined clash of civilizations.* Göteborg: Nordicom.

Engesser, S., & Brüggemann, M. (2015). Mapping the minds of the mediators: The cognitive frames of climate journalists from five countries. *Public Understanding of Science,* 1–17. doi:10.1177/0963662515583621.

Fraser, N. (2014). Transnationalizing the public sphere: On the legitimacy and efficiency of public opinion in a post-Westphalian world. In K. Nash (Ed.), *Transnationalizing the public sphere* (pp. 8–42). Cambridge: Polity.

Hallin, D. C., & Mancini, P. (2004). *Comparing media systems: Three models of media and politics.* New York: Cambridge University Press.

Hanitzsch, T. (2007). Deconstructing journalism culture: Toward a universal theory. *Communication Theory, 17*(4), 367–385.

Hermida, A. (2010). Twittering the news: The emergence of ambient journalism. *Journalism Practice, 4*(3), 297–308.

Hiles, S. S., & Hinnant, A. (2014). Climate change in the newsroom: Journalists' evolving standards of objectivity when covering global warming. *Science Communication, 36*(4), 428–453.

Honneth, A. (2014). *Freedoms right. The social foundations of democratic life.* Cambridge: Polity.

Jackson, J. A. (Ed.). (2010). *Professions and professionalization: Volume 3, sociological studies.* Cambridge: Cambridge University Press.

Kunelius, R., Eide, E., Hahn, O., & Schroeder, R. (Eds.). (2007). *Reading the Mohammed Cartoons Controversy.* Bochum/Freiburg: Projekt Verlag

Larson, M. S. (1978). *The rise of professionalism: A sociological analysis.* Berkeley, CA: University of California Press.

Lewis, S. C. (2012). The tension between professional control and open participation. Journalism and its boundaries. *Information, Communication and Society, 15*(6), 836–866.

Lynch, J., & McGoldrick, A. (2005). *Peace journalism.* Stroud: Hawthorn Press.

Nerone, J. (2015). *The media and public life: A history.* Cambridge: Polity.

Nossek, H. (2004). Our news and their news: The role of national identity in the coverage of foreign news. *Journalism, 5*(3), 343–368.

Painter, J. (2013). *Climate change in the media: Reporting risk and uncertainty.* Oxford, UK: I.B. Tauris and Reuters Institute for the Study of Journalism.

Painter, J. (2015b). The effectiveness of the IPCC communication: A survey of (mainly) UK-based users. *Advance paper submitted to the IPCC Expert Meeting on Communication.* IPCC, Oslo, Norway. Retrieved from http://www.ipcc.ch/meeting_documentation/meeting_documentation_ipcc_workshops_and_expert_meetings.shtml

Pearce, F. (2010). *The climate files: The battle for the truth about global warming.* London: Guardian Books.

Powers, M. (2016). NGO Publicity and Reinforcing Path Dependencies: Explaining the Persistence of Media-Centered Publicity Strategies. *International Journal of Press and Politics.* July 2016 DOI: 10.1177/1940161216658373, pp. 1–16.

Reese, S. (2008). Theorizing a globalized journalism. In M. Löffelholz & D. Weaver (Eds.), *Global journalism research: Theories, methods, findings, future* (pp. 240–252). Malden, MA: Blackwell.

Rosen, J. (1999). *What are journalists for?* New Haven: Yale University Press.

Russell, A. (2013). Innovation in hybrid spaces: 2011 UN climate summit and the expanding journalism landscape. *Journalism: Theory, Practice & Criticism, 14*(7), 904–920.

Russell, A. (2016). *Journalism as activism: Recording media power.* Cambridge: Polity.

Secko, D., Amend, E., & Friday, T. (2013). Four models of science journalism: A synthesis and practical assessment. *Journalism Practice, 7*(1), 62–80.

Splichal, S. (2009). "New" media, "old" theories: Does the (national) public melt into the air of global governance? *European Journal of Communication, 24*(4), 391–405.

Tuchman, G. (1978). *Making news.* New York: The Free Press.

Usher, N. (2014). *Making news at the New York Times.* Ann Arbor, MI: University of Michigan Press.

Waisbord, S. (2013). *Reinventing professionalism: Journalism and the news in global perspective.* Cambridge: Polity Press.

Zelizer, B. (1993). Journalists as interpretive communities. *Critical Studies in Media Communication, 10*(3), 219–237.

Zelizer, B. (2010). Journalists as interpretive communities revisited. In S. Allan (Ed.), *The Routledge companion to journalism and the news* (pp. 181–190). New York, NY: Routledge.

CHAPTER 13

Conclusion: From Assessments to Solutions

Elisabeth Eide, Risto Kunelius, Matthew Tegelberg,
and Dmitry Yagodin

To begin with, it is an understatement to say that climate change is a complex global problem. As we start to tackle the challenges of *communicating,* these complexities increase. There is no privileged vantage point from which a clear assessment and conclusions about climate change communication in general and the IPCC in particular can be reached. Like general climate science, communication and media research can, at best, only help to provide resources for thinking about how to tackle these challenges.

This book is based on the premise that a key role of researchers is to provide diverse resources for public deliberation, and that when we are rec-

E. Eide (✉)
Department of Journalism and Media Studies, Oslo and Akershus University
College of Applied Sciences, Oslo, Norway

R. Kunelius
School of Social Sciences and Humanities, University of Tampere,
Tampere, Finland

M. Tegelberg
Department of Social Science, York University, Toronto, ON, Canada

D. Yagodin
School of Communication, Media and Theatre, University of Tampere,
Tampere, Finland

© The Author(s) 2017   281
R. Kunelius et al. (eds.), *Media and Global Climate Knowledge,*
DOI 10.1057/978-1-137-52321-1_13

ognizing increasingly global problems, global diversity must be built into the production of scientific knowledge. Hence the study at hand is the joint product of a *network* of researchers, working in different locations, cultured in diverse academic traditions and situated in a variety of institutional settings. In a modest way, by drawing on our situated experiences and expertise, by thinking through this evidence together, we have tried to do for (IPCC) climate communication something parallel to what climate scientists must do, in an exponentially more complex sense, for the whole phenomenon of climate change. They collect diverse evidence globally, exchange and compare materials, combine schools of thought, re-think and learn from each other and from mistakes, and finally synthesize the "results" into a reasonably coherent report. Even within the relatively "narrow" field of media and communication research, experiencing the challenges of synchronizing such work makes us appreciate the complications that an organization like the IPCC faces in the interfaces between the global and the local, between science and politics, the present and the future.

Facing complexity in research, it is as crucial to embrace diverse evidence as it is to get things right. However, facing severe and high-risk problems such as climate change, it is also crucial to be able to draw conclusions and think forward. What should be done? For the IPCC, this question was addressed in October 2015 by its freshly elected chair, Korean economist Hoesung Lee. In an interview with *Nature*, his views suggest that the panel is turning a corner: placing more emphasis on social science, initiating more interaction with business and finance, and most of all, paying more attention to *solutions* for climate change.[1] Reflecting on the road ahead, the same issue of *Nature* effectively outlines four challenges: calling for more *agility* (policymakers need shorter and more focused thematic reports), *interdisciplinarity* (policymakers must better understand how climate information is trusted and used, drawing more from the social sciences and humanities), *inclusivity* (scientific networks must be strengthened in countries likely to suffer hard) and *comprehensibility* (the panel's findings must be communicated better to non-scientific audiences, i.e. most citizens).[2] All these challenges are intimately entangled in *ques-*

[1] Tollefson, J. (2015). Climate-panel chief Hoesung Lee wants focus on solutions: The newly elected chairman of the Intergovernmental Panel on Climate Change outlines his philosophy. *Nature*, October 13. Retrieved from http://www.nature.com/news/climate-panel-chief-hoesung-lee-wants-focus-on-solutions-1.18556

[2] Schiermeier, Q., & Tollefson, J. (2015). Four challenges facing newly elected climate chief: New leader of Intergovernmental Panel on Climate Change, Hoesung Lee, has a full in-tray. *Nature*, October 6. Retrieved from http://www.nature.com/news/four-challenges-facing-newly-elected-climate-chief-1.18492

tions of communication. Moving toward solving them demands a better understanding of the evolving *transnational communication infrastructure*—the interplay of contexts, institutions and networks—that are in play in the mediation of climate knowledge.

Journalism is one key element in this infrastructure that shapes the way IPCC knowledge and results are circulated. Our research has tried to capture a broad picture of the global landscape in which the dissemination and re-interpretation of the IPCC results is situated. By looking at the coverage, in several countries across all corners of the world, and by listening to key scientists and journalists involved in this process, we aimed at mapping the dynamics of this infrastructure.

KEY FINDINGS

Mainstream journalism, of course, is not the only channel through which IPCC influence is transmitted. Professional networks of scientists, civil society activists and NGOs as well as political and business networks and institutions cut across national boundaries. These actors often convey, translate (and twist) the impact of "global" IPCC knowledge. Yet by keeping up national, and to a lesser degree, transnational spaces of interpretation, journalism and the media are one key element in the environment of these other climate policy actors. Our research highlights many of the challenges and obstacles that the nature of the current, journalistic infrastructure poses to lively and robust transnational climate communication. While one study cannot cover this ground exhaustively, we have captured some of its essential features and dynamics.

Our findings suggest a particular scientific perspective and ethos dominates coverage of the IPCC report, often playing down the views of economics, social science, civil society and business actors (Chap. 2). Instead, journalists largely shaped the public spaces of interpretation by drawing on the IPCC's unified natural scientific evidence (Chap. 3). They also introduce this evidence through a prism of national political agendas, relying on interpretations by domestic politicians while less seldom turning to other sectors for more social, cultural or economically oriented explanations. We also found—unsurprisingly but not less meaningfully—that the structures of global communication flows led to an uneven distribution of global scientific knowledge: social (HDI) and economic (GDP) development indicators are connected to the quality of science communication in different countries, suggesting that social and economic inequalities are reflected in knowledge and communication deficiencies. As for the interplay of science and journalism, various socio-economic indicators divide the global communication space into four distinctive types of science–journalism–policy relation (Chap. 4). Thematic and conceptual comparisons add flesh to these findings. Our work points to how particular themes, relating to climate change and climate science, dominated coverage. Overall, journalism still commonly emphasized the disaster theme while crucial value frames, such as global justice or solidarity, were relatively weakly developed and (again) unevenly distributed (Chaps. 5 and 6).

In order to highlight and reflect on the particular factors in play, we looked at groupings of countries with a more specific focus in mind. Comparing four emerging economies (Brazil, Russia, China and South Africa) we saw that despite a broad and shared national *economic interest* and position on the right to economic growth, the IPCC topics received quite different interpretations. Economic scenarios ranged from an abstract reformulation of global economic principles (Brazil and Russia) to more concrete practical adjustments to industrial development (China and South Africa) (Chap. 7). For a more in-depth look at the relationship, a three country comparison of the role of policy-networks in energy politics (Australia, France and Japan) suggests a common feature: the parallel coexistence of energy-driven strategic policy communities and more alternative economic initiatives or environmental groups. What we find lacking is the ability of journalism to effectively contribute to the task of connecting these networks with local political systems. By partially shaping the environmental dynamics in which policy actors work, journalism influences their strategies, options and choices. This clearly needs to be studied more (Chap. 8). Another take on *networks* of communication looked at how social media and its networking capacities supplemented and diversified the IPCC's communication efforts. Adapting complex science messages to the formats on Twitter opened the IPCC findings up to wider and faster circulation across all kinds of institutional and social landscapes. This effort put the power of traditional journalism to the test and highlighted the growing importance of new types of professional communicators (Chap. 9).

We also took a more in-depth look at the often under-researched contexts in developing countries (Bangladesh, Chile, Egypt, Ethiopia, Indonesia and Uganda). What remains obvious is that despite the IPCC's efforts to broaden the scope of science representation from the Global South, media coverage in these countries remains highly influenced by views generated in the Global North. The lesson of inclusivity must be stressed. A bottom-up approach to climate science communication—paying more attention to victims of dramatic environmental changes, personalizing coverage and promoting a message of hope in developing countries—is a vital part of any successful effort to mediate climate knowledge (Chap. 10). This is not only a question of more effective distribution of global assessments (IPCC) to different localities but also, crucially, a question of supporting the translation of local experience and knowledge to influence global debates. As a contribution to the practical search for how journalism can become part of this effort, we presented examples of particularly good journalistic work found in the coverage. This analysis again emphasized the need for a closer relationship between climate scientists and journalists. In many contexts, the obstacles for better practices of communicating climate science include a lack of editorial resources and poor journalistic education on science-related knowledge. Solutions included professional specialization that may lead to a more distinct form of climate science journalism (Chap. 11). The issue of specialization and the professional role of journalism led us to look at the professional roles, values and identities that informed the work of some of the key journalists in the global IPCC AR5 coverage. What we found was a mixture of traditional professional values and some sharp criticism of the IPCC communication effort. But we also found an interesting interpretative community at the intersection of science, climate change and journalism: journalists with a high commitment and considerable affinity to the scientific community; professionals adapting to new roles and relationships with climate scientists; and policymakers experimenting with new modes of information production and dissemination (Chap. 12).

Lessons for the IPCC

Focusing on the intersection between the IPCC and (global) journalism offers a snapshot of the transnational infrastructure of climate communication and some of its key dynamics during 2013–2014. This analysis, we hope, can serve as a moment of mutual enrichment. On the one hand, a better understanding of the rules, routines and repertoires of journalism can help the IPCC and the broader community of transnational climate governance to communicate more constructively. As we write these conclusions, the IPCC has already begun to take journalistic and other communication challenges more seriously. In February 2016, an IPCC expert meeting on the AR5 communication effort concluded that this "integral part of the report process and work on communications [...] should start at the outset of developing the assessment" (IPCC; Lynn et al. 2016: p. 46). Among the recommendations were engagement with stakeholders from the outset, and staying more conscious of the changing media/technology landscape. Not least, they realize that "Communicating IPCC findings to diverse audiences is a huge task. The IPCC cannot do it all. Third parties have an important role to play and the IPCC must define how it will work with them" (ibid.)[3] The report also recommends more training for IPCC authors, as well as taking on board journalists and science writers, to improve the panel's communication skills. The meeting even discussed the idea to have a separate "Summary for Citizens" but ended favoring improvement of the SPMs instead (ibid.: p. 47). This apparently stems from the realization (what some of our interviewees also suspected) that, while formally the target group, many politicians do not even read the SPMs. Thus, one of the conclusions was that the SPMs should "give priority to policy-relevant questions ... The SPM does not need to cover the findings of every chapter. It should be as short as possible, with page limits" (ibid.: p. 47).

These recommendations may also be viewed as a response to criticism raised by a research collective that conducted a linguistic analysis of the IPCC SPMs. The authors state that these "clearly stand out in terms of low readability, which has remained relatively constant despite the IPCC's efforts to consolidate and readjust its communications policy" (Barkemeyer et al. 2015: p. 311). They contrast this with newspaper coverage, which they claim "has become increasingly readable and emotive."

Whether this work, with all its recommendations, will generate substantial changes in the IPCC's communications strategy remains to be seen. What this demonstrates, however, is a growing recognition that as "solutions" become a more central focus of the IPCC's work, framing communication issues as a challenge of "outreach" or "dissemination" may eventually become too narrowly focused. If business and civil society actors are to be more integrated into the work of the IPCC, questions concerning the role of journalism and public discourse *in the production* of IPCC knowledge will become even more acute (see also Chap. 2). How to develop the IPCC output into a more widely accessible and useful public resource is a key question for providing better climate governance locally and globally.

---

[3] The highlighting of "third parties" may be interpreted both as businesses and other parts of civil society such as NGOs, educators and journalists.

LESSONS FOR JOURNALISM

Just as scientists and political actors (in the IPCC) must incorporate a new understanding of the changing dynamics of the global communication infrastructure into their work, journalists working on climate change and science must consider how climate change *as a topic* challenges journalism as an institution. As we suggested at the beginning of this book, climate change and science exemplify the new scale and complexity of the intensively interconnected world in which journalism operates. In this world, earlier key professional distinctions—such as the relationship between knowledge and values, past and present, journalists and sources, journalism and the state—are gradually being re-negotiated. In turn, then, there are lessons that climate change can teach journalists.

The increased IPCC consensus—verified yet again in AR5—should direct the attention of journalists away from the *general*, epistemological question (Are we sure climate change is happening?) toward the *particular* question of describing the change. In many ways, the general frame of "global concern" (alarm, disaster, fear) has been important and useful, but journalism now needs to move further beyond the mission of raising general consciousness to the challenges involved in mitigation and adaptation. This is already happening, but we foresee—in the current situation of media downsizing—struggles within the newsrooms to uphold these priorities. There is a need for journalists to more effectively frame climate change on a scale of actual policy choices and with real consequences for people "on the ground." The IPCC's global knowledge needs active local translation into critical questions directed at local stakeholders, trying to bridge the gaps between politics, business and civic activism.

Journalism is always structurally dependent on the local news and their attempt to take part in the public debates that domesticate climate change and politics. Partly, these structures are beyond the control of journalists, but partly also a resource that journalism constructs and reproduces daily. Thus, keeping a more diverse space of interpretation is in the interest of journalists and their professional autonomy. Given all the local structures and mechanisms of control that restrain journalism, global issues, such as climate change, offer new opportunities based on transnational networks of actors. Forging new kinds of alliances—between media outlets, between individual journalists, between journalists and other stakeholders—can help journalists to produce more diversified spaces of public interpretation. While achieving domestic or local diversity remains a challenge in itself, global communication infrastructure could provide many more opportunities to use transnational actors and networks as a resource for critical, investigative journalism. Several recent media events pertaining to information leaks, for instance, show the exceptional potential of such transnational alliances (e.g. the ongoing debate on digital surveillance, security and privacy, or the spring 2016 Panama Papers revelations concerning elite tax havens). At a more broad and sustained level, the possibilities of integrating more global voices into local reporting are still grossly underused. They could also energize the ethical aspects of climate journalism by sharing everyday experiences. These, in turn, might help to articulate issues of justice and solidarity currently lacking in climate communication.

Transnational networks acquired special importance after the Paris Agreement was reached at the end of COP21. Producing the agreement is an achievement in itself in which the interplay between journalism and the IPCC, in communicating climate urgency, has been essential. Now the implementation becomes even more crucial as the core mechanism in the Paris Agreement is the *monitoring* of national emissions targets. The role of journalism both to provide platforms for debate and to support making goals and targets globally transparent will become extremely important.

Journalism in countries with low media attention to climate change holds the potential to explore, cultivate and promote this underdeveloped area of work. It needs to assert more consciously the environmental agenda within the newsroom hierarchies and learn to link coverage of climate science to issues that are already locally central and newsworthy. Hypothetically, low coverage also means there is more unoccupied professional space that can be populated by newcomers to journalism from the realms of science, social movements and politics.

As in the case of arguing for the IPCC to change its way, these recommendations to journalism expose not only what journalism is capable of, but also the obstacles it faces moving forward. There are severe structural limits to the work of journalists. Some are political and power-driven, such as the attention biases and dominant elite frames that emerge in one-sided local political networks. Others have to do with limited newsroom resources, issues of education and broader social structures. Thus, good climate journalism requires encouragement and support.

Lessons for (Media) Research

By listing lessons and recommendations to journalists and scientists, we are not suggesting a tug of war. Contrary to this, our work suggests the need—and the possibilities—for increased dialogue between scientists and journalists in an atmosphere of mutual recognition and confidence. Having pointed out that climate scientists and science communicators face immense common challenges in shifting toward a more "solution oriented" perspective we wish to conclude by acknowledging that for communication, media and journalism research this terrain also presents new tasks.

There is an urgent need for further cross-country studies of climate communication. This research must move away from the prevailing North/West bias toward more *dialogic* research designs. We also need to move from general "climate change" comparisons to comparisons of more particular policy problems (energy, technology, education, land-use, etc.) and their links to public communication.

Clearly, there is a need for research that spans over the diverse modes of address in the current media landscape. Here, our own approach through mainstream journalism might be justified as an entry point but it is clearly insufficient for a generalizable diagnosis of the transnational communication infrastructure that global climate governance needs. In particular, more research is needed on social media and science communication, and especially on how climate scientists can engage "directly" in dialogues with other stakeholders through social media. Indeed, social media may be harnessed to contribute to an increased awareness of "good practices." The role of transnational scientific platforms that aggregate and synthesize findings through analysis and interpretation needs to be better understood. While we already know much about the institutional practices and biases of traditional journalists and media, a better understanding of such intermediary agents of climate science and politics is crucial. These institutions perform a crucial function as an information infrastructure for politics and citizens.

Despite the growing importance and multiplicity of networked communication channels, we must still look carefully at professional journalism since it continues to play a vital role in articulating the "public"—both to itself (nationally and globally) and to the political system in the emerging hybrid media environment. The professional reflexivity and creativity of journalists, working on the "climate beat," remains an essential nodal point in this new system. A transnational dialogue on the normative and practical challenges of "solution oriented" climate journalism is a necessary element in supporting the relatively critical autonomy of journalists. Journalists and media researchers can and must be a useful element in this, not least through more emphasis on exchanging experiences. Critical readings of earlier reform movements (peace journalism, development journalism), exchanges of "good practices," and the building of professional networks are common tasks for media researchers and professionals.

Media research can also play a part in bridging communication gaps between IPCC science communicators and journalism. It is important to bring the social sciences and humanities—and, from our point of view, media researchers in particular—into a more intense and "solution oriented" conversation with climate scientists, and with the IPCC. There is a need for more experimental research that tackles the challenges of communicating different kinds of climate science (from physics and geology to economics and sociology) to groups of stakeholders. We would also benefit from more detailed analyses on cases where science communication has successfully been part of the production of informed public policy (acid rain, CFCs, greenhouse gas, tobacco, etc.). This would enhance our understanding of how to face the differentiated policy challenges of mitigating and adapting to climate change.

The chapters in this volume demonstrate convincingly that communication challenges related to climate change in general or to the IPCC in particular cannot be solved by pointing to particular actors or institutions. If climate change is, as we argued in the beginning of the book, a "post-normal" or "post-modern" problem, it follows that modern institutions—science, politics, media, etc.—must reconfigure their relationships. Thus, following Hoesung Lee's idea, an increased emphasis on "solutions" may set a good new course for the IPCC and for journalism. Moving to the direction of a more solutions-oriented IPCC will not be a simple path. In our interview with IPCC scientist Karen O'Brien (Chap. 2), for one, she pointed out that the structure of the panel (WGs) and its future reports will remain the same, with the work continuing to be guided by an economic-rationalist approach. A serious emphasis on "solutions" will pose many new challenges and obstacles, and a new forward-looking climate communication will demand new kinds of reflexivity from journalists and scientists, especially, if they are to call attention to the gaps and dysfunctionalities inherent in current practices and systems. Several of the IPCC authors who shared their reflections with us in this volume, for instance, emphasized their obligations both as scientists and as citizens. On the one hand, they underscored their responsibility to communicate in a way that makes climate change relatable to citizens. On the other hand, they recognized that they may not be able to fulfill these obligations, partly because of the complexity of the subject matter but also due to their assigned roles as merely "policy relevant" and not "policy prescriptive." A move toward knowledge about "solutions" will demand some reconsideration of this imagined borderline between science and politics. It might be highly *policy relevant* to undertake in more studies of *successful examples of adaptation and mitigation,* prioritizing such issues in forthcoming special reports.

We close by borrowing vocabulary from the IPCC. In referring to the forthcoming global warming, the panel mentions that the world has already "committed" itself to climate change. There is no other option left than to be committed to finding ways of coping with the challenges this brings to scientists, journalists and all other stakeholders.

BIBLIOGRAPHY

Barkemeyer, R., Dessai, S., Monge-Sanz, B., Gabriella Renzi, B., & Napolitano, G. (2015). Linguistic analysis of IPCC summaries for policymakers and associated coverage. *Nature Climate Change, 6,* 311–316.

IPCC (2016, February 9–10). Lynn, J., Araya, M., Christophersen, Ø., El Gizouli, I., Hassol, S.J., Konstantinidis, E.M., Mach, K.J., Meyer, L.A., Tanabe, K., Tignor, M., Tshikalanke, R., van Ypersele, J.P. (Eds.) Meeting Report of the Intergovernmental Panel on Climate Change Expert Meeting on Communication. IPCC, Geneva, Switzerland. Retrieved from http://www.ipcc.ch/meeting_documentation/meeting_documentation_ipcc_workshops_and_expert_meetings.shtml

APPENDIX: NEWSPAPER ARTICLES CODEBOOK FOR IPCC AR5 STORIES

Every newspaper item (including infographics and fact boxes) with an individual headline is coded separately.

1. **Name of the newspaper**
2. **Date of publication**
3. **Main headline of the story translated into English**
4. **Presence of the following frames in the headline (yes/no)**

(NOTE: Coded from translated headlines in Tampere)

4.1 *Global concern*—headlines that frame the issue as a problem that the whole world must face. The consequences of climate change are seen as wide reaching.

4.2 *Local/specific concern*—translated into concrete problems and challenges (geographically or thematically specified consequences, extreme weather events).

4.3 *Scientific evidence*—that does not link to specific consequences or a need to act. Includes evidence that the IPCC has produced, discussions of uncertainty and blunt factual statements or conclusions about man-made climate change.

4.4 *Politics/politicization*—emphasizes differences, controversies and skepticism. This frame includes references to politicians and their responsibility, disagreements between stakeholders and political influence of the IPCC.

© The Author(s) 2017   293
R. Kunelius et al. (eds.), *Media and Global Climate Knowledge*,
DOI 10.1057/978-1-137-52321-1

4.5   *General call for action*—demanding prevention of climate change (cutting emissions generally, without concrete targets), its mitigation or better adaptation without references to specific geographical areas or specific field of action (energy, food supply).

4.6   *Specified call for action*—identifies the need for more concrete policy areas or measures, sometimes focusing on specific geographic areas or specific climate policy targets.

4.7   *Moral references*—to justice, responsibility or other moral arguments. Emphasis on fairness and responsibility in relation to time (future generations), location (south–north), power hierarchies (politicians–people, developed–developing world); explicit use of "we" in demands of action and references to "blame".

5. **Name of the journalist**
6. **Gender of the journalist (female, male, both, not known)**

1   Female
2   Male
3   Both
4   Not known

7. **Story relevance (identifying different sub-samples)**

1   IPCC AR5 report is the main story
2   General Climate Change, with reference IPCC AR5
3   General climate story, no reference to IPCC AR5
4   Other main topic, reference to either IPCC or climate

If coded 4, only the part relevant to climate change or the report is coded.

8. **Story size**

1   Small (less than 300 words)
2   Medium (300–600 words)
3   Major (more than 600 words)

9. **Genre**

1   *Reporting genres* (news, reportage, interviews, profiles, background pieces, graphs and fact boxes, other reporting genres)

2   *External opinionated genres* (column or op-ed by an outside writer, letters to the editor)
3   *Internal opinionated genres* (editorial, column or comment by a staff journalist)
4   Other

10. Voices (from 0 to 15 voices possible in each story)

In letters to the editor or op-ed stories, the author is coded as the first voice, and voices quoted by the author as additional voices. In addition to coding all voices in terms of their role and competence, we distinguish between voices that are *explicitly* linked to the procedures related to the IPCC report (e.g. national representatives in negotiating and accepting the final text, working group authors or participants).

National political system (representative of a national party or institution)
11   Domestic national political actor (presidents, members of parliaments, ministers etc.)
111   Domestic national political actor (with IPCC connection)
12   Foreign national political actor
121   Foreign national political actor (with IPCC connection)
13   Other national political system

Transnational political system
21   Regional institution (EU, ASEAN, NAFTA, AU, OIC, League of Arab States, etc.)
211   Regional institution (with IPCC connection)
22   United Nations and its sub-organizations
221   United Nations and its sub-organizations (with IPCC connection)
23   Other transnational political system
231   Other transnational political system (with IPCC connection)

Civil society
31   Domestic NGO
32   Foreign NGO
33   Think tank
34   Domestic grassroots actor ("ordinary people")
35   Foreign grassroots actor ("ordinary people")
36   Other civil society

Business
41   Domestic industry or business
42   Foreign industry or business

43   Other business

Scientists and other experts

51   Domestic scientist or expert (technology)

511   Domestic scientist or expert (technology) (with IPCC connection)

52   Domestic scientist or expert (natural science)

521   Domestic scientist or expert (natural science) (with IPCC connection)

53   Domestic scientist or expert (social science and humanities)

531   Domestic scientist or expert (social science and humanities) (with IPCC connection)

54   Domestic scientist or expert (economics)

541   Domestic scientist or expert (economics) (with IPCC connection)

55   Domestic Scientist or expert (other than those mentioned above or field unknown)

551   Domestic Scientist or expert (other than those mentioned above) (with IPCC connection)

56   Foreign scientist or expert (technology)

561   Foreign scientist or expert (technology) (with IPCC connection)

57   Foreign scientist or expert (natural science)

571   Foreign scientist or expert (natural science) (with IPCC connection)

58   Foreign scientist or expert (social science and humanities)

581   Foreign scientist or expert (social science and humanities) (with IPCC connection)

59   Foreign scientist or expert (economics)

591   Foreign scientist or expert (economics) (with IPCC connection)

60   Foreign scientist or expert (other than those mentioned above)

601   Foreign scientist or expert (other than those mentioned above) (with IPCC connection)

Media (journalists)

71   Media

All others not coded above

81

10. Names of the voices quoted above and institutions they represent (as mentioned in the text)

11. Voice gender

Number of female voices

Number of male voices

BIBLIOGRAPHY

Antilla, L. (2010). Self-censorship and science: A geographical review of media coverage of climate tipping points. *Public Understanding of Science, 19*(2), 240–256.

Auer, M. R., Zhang, Y., & Lee, P. (2014). The potential of microblogs for the study of public perceptions of climate change. *Wiley Interdisciplinary Reviews: Climate Change, 5*(3), 291–296.

Banda, F. (2007). An appraisal of the application of development journalism in the context of public service broadcasting (PSB). *Communicatio: South African Journal for Communication Research and Theory, 33*(2), 154–170.

Beck, U. (2003). Toward a new critical theory with a cosmopolitan intent. *Constellations, 10*(4), 453–468.

Bennett, W. L., & Segerberg, A. (2011). Digital media and the personalization of collective action: Social technology and the organization of protests against the global economic crisis. *Information, Communication & Society, 14*(6), 770–799.

Bosch, T. (2012). Blogging and tweeting climate change in South Africa. *Ecquid Novi: African Journalism Studies, 33*(1), 44–53.

Brüggemann, M., & Wessler, H. (2014). Transnational communication as deliberation, ritual, and strategy. *Communication Theory, 24*, 394–414.

Carey, J. W. (1969). The communications revolutions and the rise of the professional communicator. In P. Halmos (Ed.), *The sociology of mass media communicators* (pp. 23–38). Keele: University of Keele.

Carey, J. W. (1997). Afterword: The culture in question. In M. E. Stryker & C. Warren (Eds.), *James Carey: A critical reader* (pp. 308–339). Minneapolis: University of Minnesota Press.

Castells, M. (2009). *Communication power*. Oxford: Oxford University Press.

Climate Barometer. (2015). *Finns want more effective measures for mitigating climate change. Summary of the survey*. Released by the Ministry of the Environment, Finland. Retrieved from http://www.ym.fi/download/noname/%7B22C22786-B04F-464B-8640-87DE9349C365%7D/108389

Cox, R. (2013). Climate change, media convergence and public uncertainty. In L. Lester & B. Hutchins (Eds.), *Environmental conflict and the media* (pp. 231–245). New York: Peter Lang.

Denham, B. E. (2010). Toward conceptual consistency in studies of agenda-building processes: A scholarly review. *Review of Communication, 10*(4), 306–323.

Djerf-Pierre, M. (2013). Green metacycles of attention: Reassessing the attention cycles of reporting 1961–2010. *Public Understanding of Science, 22*(4), 495–512.

Eide, E. (2012a). An editorial that shook the world: Global solidarity vs. editorial autonomy. In E. Eide & R. Kunelius (Eds.), *Media meets climate: The global challenge for journalism* (pp. 125–143). Gothenburg: Nordicom.

Eide, E. (2012b). Visualizing a global crisis: Constructing climate, future and present. *Conflict and Communication Online, 11*(2). Retrieved from http://www.cco.regener-online.de/2012_2/abstr_engl/eide_abstr_engl.htm

Ekecrantz, J., & Olsson, T. (1994). *Det redigerade samhället*. Stockholm: Carlssons.

Elgesem, D., Steskal, L., & Diakopoulos, N. (2015). Structural content of the discourse on climate change in the blogosphere: The big picture. *Environmental Communication, 9*(2), 169–188.

Field, C. (2014). *IPCC plenary opening session*. Geneva, Switzerland: IPCC. Retrieved from http://www.ipcc.ch/pdf/ar5/Chris%20Field%20Opening%20Statement.pdf

Field, C., Mach, K., Sokona, Y., & Stocker, T. (2016). IPCC communications: Experiences from the AR5. *Advance paper submitted to the IPCC Expert Meeting on Communication*. IPCC, Oslo, Norway. Retrieved from https://ipcc.ch/meeting_documentation/pdf/Communication/160119_advance_paper_on_constraints-JLynn.pdf

Galtung, J. (2002). Peace journalism—A challenge. In W. Kempf & H. Luostarinen (Eds.), *Journalism and the new world order: Volume 2, studying the war and the media* (pp. 259–227). Gothenburg: Nordicom.

Greenberg, J., & MacAulay, M. (2009). NPO 2.0? Exploring the web presence of environmental nonprofit organizations in Canada. *Global Media Journal: Canadian Edition, 2*(1), 63–88.

Hanitzsch, T. (2007). Deconstructing journalism culture: Toward a universal theory. *Communication Theory, 17*(4), 367–385.

Hassol, S. J. (2016). Communication in other assessments—The U.S. National Climate Assessment, 2014. *Advance paper submitted to the IPCC Expert*

Meeting on Communication. IPCC, Oslo, Norway. Retrieved from https://ipcc.ch/meeting_documentation/pdf/Communication/160119_advance_paper_on_constraints-JLynn.pdf

Hawkins, E., Edwards, T., & McNeall, T. (2014). Pause for thought. *Nature Climate Change, 4*, 154–155.

Høyer, S., & Pöttker, H. (Eds.). (2005). *Diffusion of the news paradigm 1850–2000*. Gothenburg: Nordicom.

Huq, S. (2015, July). *Keynote address. Our common future under climate change*. Paris, France: International Science Conference.

IPCC. (2016, February 9–10). Lynn, J., Araya, M., Christophersen, Ø., El Gizouli, I., Hassol, S.J., Konstantinidis, E.M., Mach, K.J., Meyer, L.A., Tanabe, K., Tignor, M., Tshikalanke, R., & van Ypersele, J.P. (Eds.), *Meeting report of the Intergovernmental Panel on Climate Change expert meeting on communication*. Geneva, Switzerland: IPCC. Retrieved from http://www.ipcc.ch/meeting_documentation/meeting_documentation_ipcc_workshops_and_expert_meetings.shtml

Krøvel, R. (2012). Setting the agenda on environmental news in Norway. *Journalism Studies, 13*(2), 259–276.

Lee, C. C., Chan, J., Pan, Z., & So, C. (2000). National prisms of a global media event. In J. Curran & M. Gurewitch (Eds.), *Mass media and society* (3rd ed.). London: Edward Arnold.

Lynn, J. (2016b). Preparations for the release of AR5 and previous reports. *Advance paper submitted to the IPCC Expert Meeting on Communication*. IPCC, Oslo, Norway. Retrieved from https://ipcc.ch/meeting_documentation/pdf/Communication/160119_advance_paper_on_constraints-JLynn.pdf

Meraz, S. (2011). The fight for 'how to think': Traditional media, social networks and issue interpretation. *Journalism, 12*(1), 107–127.

NCA. (2014). *National climate assessment: U.S. global change research program*. Retrieved from http://nca2014.globalchange.gov/

Nerlich, B., Forsyth, R., & Clarke, D. (2012). Climate in the news: How differences in media discourse between the US and UK reflect national priorities. *Environmental Communication, 6*(1), 44–63.

Nohrstedt, S. A., & Ottosen, R. (2014). *New wars, new media and new war journalism: Professional and legal challenges in conflict reporting*. Göteborg: Nordicom.

Nossek, H., & Kunelius, R. (2012). News flows, global journalism and climate summits. In E. Eide & R. Kunelius (Eds.), *Media meets climate: The global challenge for journalism* (pp. 67–86). Gothenburg: Nordicom.

Olausson, U., & Berglez, P. (2014b). Media research on climate change: Where have we been and where are we heading? *Environmental Communication, 8*(2), 139–141.

Patt, A. G., & Weber, E. (2014). Perceptions and communication strategies for the many uncertainties relevant for climate policy. *WIREs Climate Change, 5*, 219–232.

Pew Research Center. (2013). *Global attitudes project, spring 2013.* Released by Pew Research Center. Retrieved from http://www.pewglobal.org/category/datasets/

Rhaman, M. (2015). COPs and climate: The practice of development journalism in Bangladesh. *Paper presented at the conference on Communication and Environment (COCE), International Environmental Communication Association,* Colorado University, Boulder, CO, June 2015.

Schäfer, M. (2015). Climate change and the media. In P. Baltes (Ed.), *International encyclopedia of the social & behavioral sciences* (pp. 853–859). Philadelphia: Elsevier.

Shehata, A., & Hopmann, D. N. (2012). Framing climate change. *Journalism Studies, 13*(2), 175–192.

SIFO. (2013). *SIFOs telefonbuss 2013.* Released by Global Challenges Foundation. Retrieved from http://globalchallenges.org//wp-content/uploads/global-challenges-sifo-september-2013.pdf

Splichal, S. (2006). In search of a strong European public sphere: Some critical observations on conceptualisations of publicness and the (European) public sphere. *Media, Culture and Society, 28*(5), 695–714.

Sreberny-Mohammadi, A., Nordenstreng, K., Stevenson, R., & Ugboajah, F. (Eds.). (1985). *Foreign news in the media: International reporting in 29 countries.* New York: United Nations.

TNS Gallup. (2015). *Klimabarometeret 2015 rapport.* Released by TNS Gallup, Norway. Retrieved from http://www.tns-gallup.no/sokeresultat?q=klimabarometer

Ungar, S. (2014). Media context and reporting opportunities of climate change: 2012 versus 1988. *Environmental Communication, 9*(2), 233–248.

Vidal, J. (2014). 'We expect catastrophe': Manila, the megacity on the climate frontline. *The Guardian,* March 31. Retrieved from http://www.theguardian.com/environment/2014/mar/31/ipcc-climate-change-cities-manila

World Risk Report. (2013). United Nations University, Institute for Environment and Human Security. Bündnis Entwicklung Hilft (Alliance Development Works). Retrieved from http://worldriskreport.entwicklung-hilft.de/uploads/media/WorldRiskReport_2013_online_01.pdf

Wosniak, A., Lück, J., & Wessler, H. (2014). Frames, stories and images: The advantages of a multimodal approach in comparative media content research on climate change. *Environmental Communication, 9*(4), 469–490.

Wübbeke, J. (2013). The science-politics of climate change in China: Development, equity, and responsibility. *Nature & Culture, 8*(1), 8–29.

Xinsheng, L., Lindquist, E., & Vedlitz, A. (2011). Explaining media and congressional attention to global climate change, 1969–2005: An empirical test of agenda-setting theory. *Political Research Quarterly, 64*(2), 405–419.

INDEX[1]

A

access, 20–7, 59–79, 83, 84, 88–92, 110, 130, 133, 155, 158, 178, 179, 221, 254

Action Aid, 205

actor-network, 25, 95, 194–202, 208
   actor-network theory (ANT), 195, 196

adaptation, 14, 22, 37, 38, 50, 53, 62, 68–70, 74, 77n7, 79, 86, 87, 89, 95, 97, 99, 100, 109, 121, 156, 160, 164, 185, 188, 214, 215, 221, 224, 227, 240–2, 244, 247, 272, 274, 287, 291, 294

advocacy, 135, 155, 178, 184, 189, 202–3, 208, 230, 257, 258

Africa, 19, 21, 38, 67, 73, 75, 86, 89, 91, 97, 98, 112, 113n1, 116, 120, 123, 124, 152, 155–60, 163, 165–7, 216, 221, 222, 245, 246, 284
   African media, 155, 157, 220, 221, 226n16

alarm, alarmism, 49, 67, 70–5, 113, 227, 227n22, 287

Alberta, 198

alternative energy sources, 165

Amazon, 131, 161

Antarctica, 207

Arctic, 111, 113, 116, 198, 239–40, 253

The Asahi Shimbun, 19, 113n1, 179, 264

Asia, 69, 142n7, 165, 216, 221, 227, 227n24, 243

Asia-Pacific Economic Cooperation, 165

attention, 7, 14, 16, 17, 20–3, 33–55, 59–79, 83–5, 88–102, 110, 115, 140, 160, 166, 172, 194, 202–4, 206, 219, 229, 242, 244, 245, 258, 268, 269, 271, 274, 282, 285, 287, 288, 290

Australia, 19, 21, 67, 86, 89, 90, 92–4, 114, 121, 132, 136, 171–90, 198, 201, 242, 264, 270, 272, 274, 284

awareness, 34, 42, 48, 86–8, 92, 95, 97, 99, 134, 140, 156, 176, 215, 248, 252, 262, 275, 289

[1] Note: Page numbers followed by "n" denote footnotes

© The Author(s) 2017
R. Kunelius et al. (eds.), *Media and Global Climate Knowledge*,
DOI 10.1057/978-1-137-52321-1

H
hashtags, 196–9
holistic, 173, 245, 253
Human Development Index (HDI),
86–9, 92, 97, 99, 101, 156, 157,
214, 215, 283

I
inclusive culture, 104
The Independent, 120, 184, 216, 269
Indonesia, 19, 21, 73, 75, 86, 90, 99,
101, 112, 114, 116, 120, 214,
215, 219, 220, 222, 223, 228,
239, 250, 253, 264, 266, 270,
272, 274, 275, 284
interdisciplinary, 14, 29
intergovernmental negotiation, 161
international news agencies, 84, 220
Internet Penetration rates, 87
interpretive community, 201, 263
Israel, 19, 21, 86, 91, 99, 101, 266
issue networks, 183–8

J
Japan, 19, 21, 86, 91, 95–7, 105, 112,
113, 116, 117, 121, 124, 136,
143, 145, 147, 171–90, 264,
266, 272, 284
journalism, journalist
(journalistic) field, 45, 85, 159, 252,
254, 277
cultures of, 84
global, 27, 53, 78, 79, 101, 102,
262, 285
professionalism, 258, 277
transnational, 15, 28, 78, 82, 105, 290
justice, injustice
recognition, 131, 142, 146
redistribution, 131, 143, 146
representation, 131, 135, 144
solidarity, 27, 129–47

K
Kompas, 19, 73, 223, 223n9, 228n27,
250, 252, 253, 264, 270, 272,
274, 275
Kyoto protocol, 157, 177

L
L'Humanité, 19, 75, 178, 179, 264,
268
lay expertise, 39, 40, 52, 55, 78
Le Figaro, 113n1, 122, 178
Le Monde, 19, 67, 113n1, 144–5, 178,
179, 187, 249, 252, 264, 270,
271
liberal model, 153, 155, 174, 178,
179
liberalization, 40
Libération, 178, 187
likelihood, 41, 42, 44, 65, 74, 120
LinkedIn, 194
lobby
fossil fuel lobby, 12, 184, 185
lobby groups, 113, 237, 274

M
Maldives, 239, 247, 248, 253
market liberalism, 84
mass media, 16, 102, 153, 154,
204–9
material semiotics, 195
media event(s), 15–17, 196, 201, 207,
288
media freedom, 83, 85–7, 89, 92, 97,
156, 157, 215
media systems, 84, 104, 153–5, 167,
174, 175
MediaClimate network, 15, 17, 17n1,
84, 237
mediated civic epistemologies, 27,
81–105
microblogging, 194

The manufacturer's authorised representative in the EU is Springer
Nature Customer Service Centre GmbH, Europaplatz 3, 69115 Heidelberg,
Germany. If you have any concerns regarding our products, please
contact ProductSafety@springernature.com

Printed and bound by CPI Group (UK) Ltd, Croydon, CR0 4YY
23/04/2026
02095601-0008